HOW TO COMPETE IN THE AGE OF ARTIFICIAL INTELLIGENCE

IMPLEMENTING A COLLABORATIVE HUMAN-MACHINE STRATEGY FOR YOUR BUSINESS

Soumendra Mohanty
Sachin Vyas

Apress®

How to Compete in the Age of Artificial Intelligence: Implementing a Collaborative Human-Machine Strategy for Your Business

Soumendra Mohanty
Kolkata, India

Sachin Vyas
Pune, India

ISBN-13 (pbk): 978-1-4842-3807-3
https://doi.org/10.1007/978-1-4842-3808-0

ISBN-13 (electronic): 978-1-4842-3808-0

Library of Congress Control Number: 2018958917

Managing Director, Apress Media LLC: Welmoed Spahr
Acquisitions Editor: Shiva Ramachandran
Development Editor: Laura Berendson
Coordinating Editor: Rita Fernando

Cover designed by eStudioCalamar

Distributed to the book trade worldwide by Springer Science+Business Media New York, 233 Spring Street, 6th Floor, New York, NY 10013. Phone 1-800-SPRINGER, fax (201) 348-4505, e-mail orders-ny@springer-sbm.com, or visit www.springeronline.com. Apress Media, LLC is a California LLC and the sole member (owner) is Springer Science + Business Media Finance Inc (SSBM Finance Inc). SSBM Finance Inc is a **Delaware** corporation.

For information on translations, please e-mail rights@apress.com, or visit http://www.apress.com/rights-permissions.

Apress titles may be purchased in bulk for academic, corporate, or promotional use. eBook versions and licenses are also available for most titles. For more information, reference our Print and eBook Bulk Sales web page at http://www.apress.com/bulk-sales.

Any source code or other supplementary material referenced by the author in this book is available to readers on GitHub via the book's product page, located at www.apress.com/9781484238073. For more detailed information, please visit http://www.apress.com/source-code.

Printed on acid-free paper

Contents

Contents

About the Authors

Soumendra Mohanty is an acclaimed thought leader and SME in the areas of analytics, IoT, AI, cognition, and automation. He is a two-decade veteran with expertise in next-gen Big Data solutions, BI architectures, the enterprise data warehouse, customer insight solutions, and industry-specific advanced analytics solutions. With his broad experience, he has designed and implemented data analytics solutions for Fortune 500 clients across industry verticals. Soumendra is an advisor with the Harvard Business Review Advisory Council. He is also associated with the Indian Statistical Institute and various universities as a visiting faculty member specializing in Big Data and analytics. Soumendra speaks at various global forums, CAO advisory forums, and educational institutions. He is author of several books including *Big Data Imperatives* (Apress).

Sachin Vyas is an entrepreneur with over 20 years of experience in technology, data, and analytics. He was the founder and CEO of AugmentIQ Data Sciences, a company acquired by LTI. LTI is a global technology consulting and digital solutions company. AugmentIQ focused on creating platforms and solutions specific to Big Data engineering and data sciences and provided solutions to large and complex data and analytics problems for financial services companies. He currently heads LTI's platforms and enabling solutions with converging technologies across devices, data, computing, and artificial intelligence. Sachin is the recipient of the EMC Transformer Awards-2012 and the SKOCH Digital Inclusion Awards-2012 in India. He has a mechanical engineering degree from VNIT, Nagpur, India.

Acknowledgments

During the writing of this book, we were fortunate to be amidst many well-wishers, colleagues, and our families, who have been immensely helpful and supportive by playing several roles—co-writers, coaches, and down-right fierce critics. This book could not have been written without their help.

We are also grateful to the many clients, analysts, and strangers who unknowingly contributed to this book when they patiently listened to our ramblings, downright stupid questions, and crazy thoughts. Sometimes they candidly admitted when something did not make sense and sometimes they vociferously agreed and nudged us to keep researching and writing.

Our efforts to write this book would have been futile without the continuous guidance from the Apress team (Shiva, Rita, and Laura). You have our sincere appreciation for your support throughout this project.

A special thank you to Rita Fernando Kim from Apress for her hard work on reviews and for keeping us on schedule.

I am deeply indebted to my loving family—Snigdha, Pratik, Pratyush, and Rexie. Without their love, affection, and support, it would be impossible to survive in the corporate madness, let alone write books.

—Soumendra Mohanty

I am grateful to Soum (Soumendra Mohanty) for the opportunity to work with him on this book. It is his idea and his book, and I am really glad that I could contribute in some ways. There couldn't have been a better opportunity for "my first-book" project! I'd also like to thank Shweta, my wife, for her patience and support throughout all my ventures, including this one.

—Sachin Vyas

Introduction

Over the last several years, as we met with hundreds of CxOs, senior executives, business function owners, technology leaders, and practitioners, we realized that there are fundamental questions that everybody was trying to make some sense of:

- What is AI?

- Why now?

- How does it impact my function or business or even the company at large?

- What should I do?

This book is collection of thoughts, assimilating ideas, and views—some thought provoking, some mundane—addressing several aspects of how to compete in the age of artificial intelligence. Once you start reading the book, you will realize that the chapters are written in a blogish manner, which is precisely by design. This is to give you free-flowing thought and help you generate your own ideas.

With the short preamble out of the way, let's get started.

The original AI-powered Watson supercomputer that conquered the contestants on the *Jeopardy* television game show was about the size of a bedroom, with 10 odd rack-like machines forming the four walls. Today's intelligent machines are very different. They no longer exist solely within bedroom sized rooms, but are cloudified and run several 100 "instances" of AI services at once. The outputs are consumed simultaneously by recipients all over the world, through multiple channels—smart assistants, smart phones, smart devices, smart appliances, smart vehicles, smart utilities, smart plants, smart factories, smart buildings, smart homes, etc. The result? We are living in a world of "always-on" intelligence.

This pervasiveness of AI is triggering another interesting phenomenon—the more we use it, the smarter it becomes. Anything it learns in one interaction gets immediately transferred to the other interactions. AI is not one large monolithic program. It is actually a combination of diverse sets of artificial narrow agents (ANIs), each specialized to do a particular task and with capabilities like conversational interface, image recognition, voice recognition, text-to-speech, speech-to-text, logic-deduction, natural language parsing, natural language generation, knowledge base, self-learning, and adaptive engines.

Slowly and steadily, a picture of the AI future is emerging. The AI on the horizon looks more like "AI as a service" embedded into everything, and almost invisible. A century ago, we transformed everything to be powered by electricity to augment human capabilities; going forward we are at a similar tipping point, where we will "AI-fy" everything to augment human capabilities and in some cases introduce autonomous AI to eliminate human tasks altogether. Thanks to technologies like Cloud, GPU, Big Data, Blockchain, IoT, ML and DL, the process of "AI-fying" will become simpler in the coming days, transforming everything by infusing it with AI. In fact, don't get surprised if the business plans of the companies going forward are going to take X and add AI.

Take the example of Google. Every time we type a query or click on a link, we are becoming an active participant in training the Google AI. Starting from how to date to how to conceive to how to raise children to DIY scientific experiments to how to manage relationships to how to prepare an effective resume to what are the most asked interview questions and corresponding answers, with each of the queries we are feeding into Google's search engine, we are helping Google AI record our behavior. This includes what we think, what we want to know, when we seek information, why we are seeking information, how we use that information, and so on. Perhaps in 10 years, Google's main product will be "Mind as a Service" and you will be able to rent the mind of a CEO, developer, or scientist.

This is the point where things become sketchy. Ethical policies and associated debates come into play.

What is propelling this massive growth of AI? Three technology breakthroughs acted as catalysts:

- **Massively parallel computation:** Thinking, which is the process of reasoning about something, is a massively parallel process where billions of neurons in our brain act simultaneously, passing signals to other neurons through layers of networks. The final outcome is judgment. We were handicapped to perform massively parallel processing until the graphics processing unit (GPU) was invented. The GPU unlocked new possibilities, where neural networks (loosely based on the way neurons work in our brain) can facilitate hundreds of millions of connections between the nodes, almost at a sub-second processing time.

- **Big Data:** The intelligence that we claim is ours is taught to us over time. When the human brain sees something it's never seen before, it takes time to deduce what the image is! The same rule applies to AI. Thanks to digitization and proliferation of smart phones, we have access to massive amounts of real-world data. Our ability to collect, clean, standardize, and store this real-world data provides us with an enormous training ground for AI. The result? We are beginning to see intelligence infused into almost everything, consequently transforming everything into a "smart" thing.

- **Better algorithms:** A lot goes on inside our brain in order to analyze our surroundings. We use heuristics and mental mind-maps to reason. However, it is incredibly hard for us to codify the thinking process. Our earlier endeavors to codify the reasoning process resulted in writing lengthy programs, mostly in the form of "IF then Else…" and these programs were not adaptive enough to changes. Deep-learning algorithms give us a way to generate reasoning from the data itself, not from complex programs consisting of hundreds and thousands of conditions, but from the data itself churning out patterns and recommendations. We can now collect lots of data and apply sophisticated algorithms to arrive at predictions. The only drawback is that some algorithms are so complex that we just can't understand the results we get.

To illustrate the impact of these three catalysts, let's discuss what happened after IBM's supercomputer Deep Blue defeated the reigning chess grand master Garry Kasparov in a famous man-versus-machine match in 1997. Kasparov had his superlative genius mind on his side; Deep Blue had instant access to a massive database of all previous chess moves played thousands of experts. Kasparov realized that man-plus-machine would be far more beneficial than man-versus-machine. The concept of a "centaur player" (a human/AI cyborg) began to emerge where AI augments the human-chess player's skills. The result? Today, the best chess player is "Intagrand"—a team of humans and several different chess AI programs.

Interestingly, another pattern began to emerge. Instead of diminishing the interest of human chess players, AI-enabled chess programs actually inspired more people than ever to play chess. Perhaps there was something really motivating and rewarding for the human chess players and the AI chess programs, both turning up the ante to stay one step ahead of the other, and consequently both learning from each other to become better players. The result? There are more grand masters now than there were when Deep Blue first defeated Kasparov.

This brings up another point. If AI can help humans become better chess players, by all means it is also possible for AI to help us become better in other spheres of life and professions.

Fantastic so far! How do all of these technology advances relate to business?

Well, the signs are there, however subtle at this point in time they may be. The world around us, and hence our businesses and the roles we play in a business context, are increasingly impacted by AI prevalence. It will require CxOs to adopt a new approach to role delegation—from implementing strategic AI advisors to augmenting employees.

Modern organizations value empowered AI as much as they value empowered people. For example, CEOs must make it clear when smart algorithms, rather than human associates, are to be consulted. This can be difficult. Some of the most important decisions regarding machine learning are usually about the extent of authority the AI agents should have. Business leaders who automate a factory now recoil at the idea of letting AI determine their entire business strategy. In the absence of clear delineation of authority as well as accountability, human-AI conflict will result from dual empowerment.

Business leaders committed to taking advantage of AI should consider the following four AI-related imperatives.

- *AI as strategic advisors:* AI can effectively play the role of the strategic advisor. The primary deliverables of the AI strategist are to reframe problems, assist in strategy development, and assist in defining the course of action and the execution plans. These algorithms will constantly produce data-driven insights and recommend optimal courses of action. Human intervention and oversight will solely be required to identify which decisions are deferred to the algorithms and how the decisions are implemented.

  The AI strategist will have a broader view across the systems and business processes and competitive intelligence, which will invariably present major operational challenges, such as inter-process and interpersonal conflicts. Hence, it is all the more important that the output of the AI strategist should be transparent and explainable to all. That way the trust factor will evolve between human and machine.

- *AI as task executioners:* The AI strategist has done its job in defining the problem statement and execution plans, and the human associate has validated it and given a go-ahead. What next? You will need algorithms to analyze business processes, create clear task descriptions and goals, and define detailed service-level agreements and key performance indicators. Managers and employees responsible for task execution will now perform higher order tasks—instead of focusing on operationalizing (allocating resources, managing employees, and managing escalations) the project execution, they will spend most of their time reviewing whether the algorithms are offering improved outcomes and innovation. Among the important benefits that algorithms offer are high-level reliability and predictability.

- *AI as a virtual assistant to employees:* Even the most talented employees have their limits. Compared to these employees, AI algorithms are geniuses boxed into a task's scope. Therefore, the question on business leaders' minds is whether the average manager and employee can effectively work together with intelligent agents.

  Modern enterprises like Google, Alibaba, Amazon, Netflix, etc. have already started using AI assistants in cases where actionable insights can improve the productivity of their employees and can help achieve the business outcomes. Employees have no other choice but to learn to treat the algorithms as valued. On the one hand, the collaboration culture of the human-machine will mean much more scalability and predictability to the outcomes. At the same time, it may to a large extent weaken accountability in the organization. One reason for this may be that in rapidly evolving scenarios, where everybody (human and machine) is running against time to deliver the outcomes, it may not be clear to managers whether they need to retrain the employee or the algorithms.

- *AI as an autonomous organization:* Is this possible? Well, if you have done the previous three, then why not? It is already happening with many Wall Street hedge funds. These companies are allowing AI full autonomy in steering the organization to new levels of risk, profitability, and innovation. Executives in these hedge funds have handed over an unimaginable portion of the decision-making process to their algorithms.

Final Thoughts

The four imperatives of incorporating AI into the enterprise may seem far-fetched now, but this process is inevitable. As the capabilities of AI continue to advance, increased oversight will lead to additional insight, and the software will learn continuously and will be viewed as an accountable agent rather than inanimate code. However, CEOs, board members, and senior executives will have a critical role to play—they must closely monitor the algorithms and promote simulations to determine the boundaries of the technology. Furthermore, business leaders should be careful about doing too many AI-led activities too soon, as doing so may create unforeseen implications on responsibility and accountability. They must ensure clarity in deference, delegation, and direction.

In this book, chapter by chapter we touch upon many aspects of AI in the context of an organization's processes, strategies, advantages, and consequences, as well as what it means for an individual in any role. In Zen, "koans" are just tools or building blocks to get to the end goal. Similarly, we sincerely hope that this book will serve as a tool to guide you and find solutions to the problems you are facing.

Now it's time to get on to the first chapter of the book and learn about the economics of AI, including what defines AI and what makes its adoption so difficult!

The Economics of Artificial Intelligence

When three types of disruptive forces (a new source of energy, a new kind of transport, and a new kind of communication) converge in concert, something dramatic happens. The cost of goods drops, the demand-supply equilibrium gets disrupted, new avenues of business opportunities open up, society by and large benefits, and we find ourselves at the cusp of an economic revolution. The last economic revolution was driven by oil (new source of energy), automobiles (new kind of transport), and telecommunication (a new kind of communication mode). In the same parlance, the new economy emerging now can be attributed to be driven by the convergence of data (the new oil), autonomous vehicles, and digital communication.

The nature of this new economy is disruptive in a massive way, as it is pivoting our current state of economics of scarcity (those having access to resources call the shots) to an economics of sharing (those who can marshal the resources call the shots). This is primarily driven by the exponential progress of technology. For example, Facebook—arguably the biggest media company in the world—has no journalists or content producers. AirBnB—the hospitality company—does not own any real-estate. Uber—the taxi company—has no cars. Cryptocurrencies—acting as banks—have no branches. All these

S. Mohanty and S. Vyas, *How to Compete in the Age of Artificial Intelligence*,
https://doi.org/10.1007/978-1-4842-3808-0_1

companies have one thing in common—they are network-based companies and are software-driven entities.

The economics of sharing abundance is all about creating opportunities where the cost to create additional products or offer a new service goes down dramatically, almost negligible tending to zero. The generation that grew up in the scarcity economy focused on hoarding and ownership was important, whereas the sharing economy is based on access to information and services, not necessarily to own them.

In a slightly different way, if the cost of production, distribution, and services tend to drop toward zero, this not only creates a sense of abundance, it also creates a sense of relative scarcity. How? One can think of this relative scarcity similar to the problem of too much information. For example, too much media (i.e. video, music, content, etc.), if made available on demand, through a preferred channel of consumption, through a device of your choice, at a fraction of a cost, can result in scarcity of attention. In short, the relative scarcity in the world of the new sharing economy will typically manifest itself in lowering our cognitive abilities—humans can't handle a barrage of new information thrown at us from all angles of life.

Our biggest challenge and our greatest opportunity lies in how we effectively and swiftly navigate this transition from economics of scarcity to economics of sharing. This is why AI (Artificial Intelligence) is a critical component in the equation. In the older economy, affordability pushed people to buy vehicles, but it created a scarcity in terms of readiness of infrastructure and roads (and therefore unbearable traffic). Now, in the newer economy, powered by AI, you will have more efficient use of vehicles (Uber), and you would see fewer ownership of vehicles. This is because owning a vehicle and bearing the brunt of traffic no longer makes sense; people rather would chose to have vehicles on demand.

Let's look at few other scenarios to see how the effects of AI play a pivotal role in the economics of sharing.

Natural Resources

Natural resources are often regulated by governments and in exchange of rights to access these resources by corporations, governments find a way to earn revenue. There are legal processes and contractual obligations to stipulate and monitor how corporations are exploiting these natural resources. These assessment methods are very primitive and are further aggravated through corrupt practices. AI and other technologies like IoT and blockchain can play a significant role in ensuring not only that resource consumption and compensation are tracked but can generate alerts on a real-time basis, wherever there are fraudulent activities manifested. In addition, all financial transactions can be made available in a public ledger that is tamper-proof and enforced at the time of consumption to eliminate any chances of corruption.

Cooperative Business

In an ideal economic setting, the producer of goods and services and the coordinator or distributor of goods and services should have a mechanism to get appropriately compensated, if not equal sharing of profits. However, in reality the value chain is extremely unbalanced, profits tend to magnify for who are responsible for coordination compared to those who provide the actual goods and services. AI can certainly play a major role in restoring some parity by providing platforms for coordination and capabilities for distribution to end consumers, thereby moving the value toward the edges to the producer of goods and services.

Energy for Everyone

Whoever has access to energy has historically enjoyed a better economy, but human greed has been consistently depleting the natural reservoirs and thus finding alternative sources of energy has become the most critical factor for governments. Solar power is quickly becoming an alternative option and the very nature of access to this source of renewable energy is poised to democratize the energy usage. AI and IoT embedded into the solar panels will allow monetization of energy, thus benefiting the general public and paving the way for efficient usage and distribution of excess energy for where it can be used more effectively.

Financial Institutions Are No Longer Intermediaries

It is almost shocking to realize that the financial industry consisting of finance, insurance, and real estate do not produce any goods or services, yet they account for a very large percentage of any nation's economy. The only thing the financial industry does is provide authenticity and commitments to the financial instruments that are required in any business transaction. The cost of this intermediation and managing the cost of financial instruments can be certainly reduced by effective usage of technologies like blockchain and AI, which in turn can greatly reduce the cost of running businesses.

Livable Cities

Over the past few decades, cities have moved away from maximizing social interactions to creating better living facilities. People are just commuting from one end to the other to go about their daily activities rather than spending quality time in a social settings and living healthily. Smart cities are hence the need of the hour, in order to provide meaningful and healthy living environments. AI and IoT technologies can deliver this objective at a fraction of the cost.

Liberated Learning

Today, one's value in the job market is determined to a large extent by a certificate from an educational institute of repute. In addition there is a distinct bias during the interview process toward the certificate over the skills and technical prowess of the candidate. Thus, it is quite natural for educational institutions to exploit this scenario. The result? The cost of education is skyrocketing and leading to crushing debt on the millennial generation. AI-based educational systems at affordable costs can greatly democratize the entire education system and can provide means to assess one's true skills almost in real time.

Truly Caring for Health

Healthcare in its current form is always a reactive practice, meaning we look for care options only when there is an emergency situation or we become sick. This behavior leads to higher monetary implications depending on the seriousness of the health issue. AI technologies can provide significant levers to make healthcare affordable. AI can lower the cost of diagnosis by analyzing an individual's past medical history and combining it with wearable devices to continuously monitor one's vital statistics, calorie intakes, physical activities, etc. This can provide alerts and recommendations driving toward preventive and proactive healthcare rather than how it is today—reactive, irreversible, and expensive.

So, how would you navigate the transition? Probably by acknowledging and prioritizing your investments and identifying which areas of your business needs reimagining with AI.

Everything Is a Prediction Problem!

The American Economic Association defines economics as "... the study of scarcity, the study of how people use resources and respond to incentives, or the study of decision-making. It often involves topics like wealth and finance, but it's not all about money. Economics is a broad discipline that helps us understand historical trends, interpret today's headlines, and make predictions about the coming years."[1]

---

[1]https://www.aeaweb.org/resources/students/what-is-economics

What is of interest in this definition are two specific points—"decision making" and "making predictions"—and these are central to what AI offers. From an economist's perspective, AI in essence is a prediction technology enabling automatic decision making, thus lowering the cost of making predictions.

What happens when the cost of prediction goes down? We have already seen the effects: prediction being a key component in determining the demand-supply equilibrium. If the cost of prediction goes down, the cost of goods and services that rely on prediction also goes down. AI dramatically drives down the cost of prediction, thereby giving impetus to two well-established economic implications. First, we start using prediction in almost everything that we do; and second, the prolific usage of prediction starts amplifying the resulting values of outcomes exponentially.

How? Consider this equation:

Outcome = Function (Data, Prediction, Decision Making, Actions)

In our daily life, whatever activities we do, can be summarized to effective usage of five components: data, prediction, judgment, action, and outcomes.

For example, I want to buy a state-of-the-art TV:

- *Gather data:* Includes make, models, brands, features, prices, discounts offered, other promotional offers, and payment options

- *Make predictions:* Such as "if I choose LG 52", then I predict outcome X, but if I chose Sony 52", then I predict outcome Y"

- *Make judgments:* "Given the features, discounts, price range, and friend's recommendations, I think the best option is Sony 52"; let me ask them to throw in a free subscription to Netflix for a year"

- *Take action:* The dealer has agreed to all that I asked for, let me swipe my credit card

- *Outcome:* I am the proud owner of a Sony 52" state-of-the-art TV, along with a year of Netflix, and an additional 15% cash back because of a promotional offer on my credit card. My wife will surely love the deal—Happy Wife, Happy Life.

As AI lowers the cost of prediction, we will begin to use it in almost everything we do, whether organizing our personal preferences or solving complex business problems. As a result, activities that were historically prediction-oriented will start becoming cheaper and better—like pricing, inventory management, logistics, supply-chain, demand forecasting, advertising, diagnosis, transportation, etc. At the same time, we will also start exploring opportunities to use prediction to solve other problems, which historically we had never even contemplated as a prediction problem.

For example, consider driving a vehicle. The common belief is that one needs to practice a lot, have a sense of direction, have a sense of traffic flow, and stay alert to react to any adverse conditions. What if we want to automate the driving experience? Humans have had a go at it and had figured out a way to develop autonomous vehicles to perform in a highly controlled environment, such as large warehouses and factories where one can anticipate a finite set of scenarios the vehicle may encounter. The key to this achievement was lengthy and complex computer programs performing if-then-else-type reasoning and instructing the vehicle to do the next best action (e.g., if there is an obstruction, then stop. Or if there is a X sign on the floor, then offload the materials there). In our wildest of dreams we had never thought to put an autonomous vehicle on a city street because the environment and interactions with the external world invites an infinite number of possible scenarios. In such an uncontrolled environment, you would need to program and implement decision making (if-then-else statements) for almost an infinite number of situations. This approach can become very costly and unmanageable, if not impossible. This method is also not fool-proof—what if you encounter a new scenario that you had never anticipated? You would have no choice but to recall your autonomous vehicle, open the program, add new conditional statements, compile the program, test it thoroughly, and then deploy the autonomous vehicle back on the streets. Definitely not a scalable method.

However, once prediction technologies became more pervasive and cheaper, AI reframed driving as a prediction problem. How? Rather than thinking through an exhaustive list of possible scenarios and then programing endless if-then-else statements, we changed the problem statement to "observe how a human driver drives" (collect lots of data), "learn from the human driver" (understand the patterns in the data), and then "predict what a human driver would do" (draw inferences and automate decision making). We installed a variety of sensors in the vehicle (inside to monitor the performance of the car, outside to monitor the external conditions), collected a huge corpus of human driving behavior, then started pattern matching by analyzing the data from outside of the car (traffic flow, traffic signals, obstructions, and proximity to other objects) to the driving decisions made by the human inside the car (navigating, steering control, braking, accelerating, turning indicators, and sometimes honking to alert other nearby vehicles or humans). In short, we made the AI learn to predict how humans react to changes in their environment. The result—self-driving cars.

Human Judgment Is Invaluable!

Prediction and judgment go hand in hand, in fact they are complementary. So, if the cost of prediction goes down, what happens to the value of judgment?

Let's take the example of a doctor examining a patient who complains about pain and swelling on his leg. First, the doctor will ask to do an x-ray of the limb and then engage in a diagnosis scenario by asking a series of questions to gather information so that she can make a prediction on what to do next. Now, AI would presumably make it easier for the doctor. By scanning through the x-ray film (image analytics) and recording the answers the patient is giving, AI can do a quick check against the huge symptoms database and then predict the best course of recommendations. However, the final decision will be left to the doctor.

So while machine intelligence could possibly substitute for human prediction, it can also be a complement to human judgment, thereby increasing the value of human judgment. In short, judgment is a complement to prediction and therefore, when the cost of prediction falls, demand for judgment rises.

When prediction becomes cheaper and cheaper, decision making becomes less cumbersome, resulting in early resolution of problems and/or uncertainties, which means we will take actions faster, which means greater demand for the application of predictions in almost anything and everything we do, which in turn means the value of judgment becomes even more valuable, which are provided by humans.

We should not generalize each and every scenario, as the line between judgment and prediction is blurry—some judgment tasks can possibly be reframed as a series of prediction tasks, each prediction feeding forward to solve the next task and so on. What is sure is that the demand for prediction-related human skills will fall, and subsequently the demand for judgment-related skills will rise.

Advancements in AI over the last decade has already achieved the status of a foundational technology component for the enterprises. It is already influencing business strategies across all dimensions—automation of business processes, transformation of customer experiences, and launching differentiated products and service offerings.

The disruptive power of AI could mean a plethora of opportunities for your business and could also mean an unsettling change that you need to manage within your business. It is therefore important to have a clear understanding of AI and understand how businesses are taking steps to drive advantage from it. In short, there is a dire need to create an AI strategy for your business.

An AI-first approach to everything also has implications, hence as business executives and technology leaders, you need to assess your business and technology landscape and then determine the appropriateness of AI-led interventions or supplements. You don't need to blindly follow what magic AI has done elsewhere. Do a lift and shift and apply to your business scenarios, this approach may do more harm. You do not want your AI transformation journey to become something new that is difficult to comprehend. You need your AI applications to be relevant to your business, you need your AI applications to take advantage of your data, and you need your AI applications to learn about and improve your past performance. And along the way, if you happen to generate new ideas that result in unique value propositions, new products, and new offerings, it is great.

AI technology is transformational and will require new leadership skills to evangelize within the enterprise. Change management is absolutely critical. The disruptive capabilities of AI are putting business leaders and technology executives under more pressure to deliver business value than ever before. The cultural change required to implement AI across the enterprise is daunting. The key driver in the AI transformation journey is not technology alone, but the focus to identify, support, and nurture the right capabilities and acquire AI talent and address competing priorities for AI investment.

The benefits of giving AI a role to play in business decision-making are many:

- **Faster decision-making**: Today every business is a digital business, thus the pace of change in business scenarios and the ability to adapt to changing conditions require businesses to speed up their decision-making process. For example, with AI-powered pricing, a business can dynamically change the price of products according to demand or competing market scenarios to improve their margins.

- **Better handling of multiple inputs**: Humans are at a disadvantage when they are presented with many factors to evaluate and make decisions. In contrast, machines can process much more data at once; they can remember much more than humans and they can use probabilistic measures to recommend the best decision to make. For example, if your business is managing logistics in the supply chain process, looking at various factors like weather conditions, fleet readiness, materials readiness, local social events, traffic conditions, etc. is important. This is all done in real time to decide on the optimal routing options. This is not something a human can comprehend, but a machine can do at scale.

- **Less decision fatigue**: It is quite understandable that if we are asked to make multiple decisions frequently and over a short period of time, our ability to deliver the most effective judgments rapidly deteriorates. In contrast, algorithms have no such decision fatigue; they will deliver the exact outcomes each time and every time. An interesting thought to ponder about—why do supermarkets have candy at cash registers? If you are spending hours in the supermarket making decisions what to buy, how much to buy, which product is better, what is on sale, etc., you are putting your thought process on super drive. By the time you get to the cashier, you will be exhausted from all the decision making. This is precisely why shoppers crave a sugar rush at the point of sale.

Given all these benefits of AI, it should be obvious that CEOs should be more than willing to hand over decision-making tasks to predictive models and algorithms! But this is not entirely true, at least as of now. Why?

Three factors act as barriers:

- **Accountability**: There is a darker side to the algorithms; they are opaque and the outcomes lack explanation. So, even if business leaders are motivated to embrace AI, they are unsure about how the algorithm arrived at the answers! It is largely a trust issue and hence in general there is reluctance to accept the AI outputs to make critical decisions.

- **Bias**: As of now, humans prepare the AI solutions, hence there is a possibility of human bias creeping into the algorithms. For example, your AI algorithm is designed to specifically filter out resumes for a job that requires STEM skills (Systems, Engineering, Mathematics, Technology) will always give preference to men. Why? It is a common notion that women are good at skills related to arts, designs, creativity, empathy, whereas men are good at skills related to logical reasoning, mental aptitude, problem solving, and doing hard things. Algorithmic bias can also happen due to data sets that are not properly curated. If the data itself is skewed to a certain inference, then the algorithm will only confirm that inference.

- **Pride**: CEOs have become leaders by going through the grueling path of management by administration, learning from their superiors, sharpening their own judgments, and being mentored by experts in the trade. In short, it takes years of learning and equal amount of gray hairs to become a CEO. Now suddenly when they have to turn to management by AI, it becomes a pride issue.

What Defines AI?

In their book *Artificial Intelligence: A Modern Approach* (Pearson, 1995), Stuart Russell and Peter Norvig defined AI as "the designing and building of intelligent agents that receive percepts from the environment and take actions that affect that environment." The most important part of this statement is the phrase "take actions." AI infuses intelligence into machines to not only learn but also to respond on its own to events from the surrounding environments at large. To respond to these events, AI needs lots of observations (data points consisting of events and corresponding actions taken) to learn from them and consequently act.

It is interesting to note that even though AI at a broad level is still a new technology, at the same time it is also going through a maturation. As soon as AI matures in one of its application areas, it gets relegated to mainstream and whatever is still maturing is branded as AI. For example, the initial experiments on AI revolved around recognizing handwriting and voices. However, with the availability of commercial systems that can decipher written text or understand human speech, these areas are no longer considered AI. Hence, arriving at a commonly agreeable definition of AI is tricky.

Some of the popular stories around AI are about beating humans at games. While playing games, humans demonstrate quick decision making, strategizing, and learning. AI not only exhibits the same human traits while playing games against them, but also becomes better than humans. While beating the humans at their own game (thinking and learning) is no doubt a commendable job, extrapolating the AI achievement from gaming situations to higher order cognitive tasks is far fetching.

First, these games have a prescribed set of rules of engagement and clear measurement of certain outcomes (e.g., win, loss, or tie). Second, these games are always played in an environment where the effect of actions is limited to participants within the system. Third, the implications of failure are no big deal; rather failure is always a fantastic learning opportunity for the AI so that it can be better trained, with no real consequences to participants outside the system.

There are currently two main schools of thought on how to develop the learning and reasoning capabilities necessary for AI programs. In both, programs learn from experience—that is, the observations and corresponding actions define the way the programs act thereafter.

The first approach uses conditional instructions (transformation rules, and heuristics)—for example, an AI program would refer to a business rules library and flag transactions as suspicious activities under the anti-money laundering context. Initially, the business rules library could have been created by looking at past evidences of what constitutes a suspicious activity and then updating continuously as newer regulations came into place or newer suspicious activities were observed.

The second approach is known as machine learning. The machine is fed with millions of observations of suspicious and non-suspicious transactions. The machine understands the various patterns in the observations leading up to a suspicious transaction. As more and more observations are fed into the AI program, the program learns and builds its understanding through this inference-making ability, without having to code specific business rules to do so.

AI applications are data-hungry and the ever increasing "datafication" by organizations, governments, households, and individuals is fueling the need to apply AI everywhere. That volume of data is due to increase exponentially on account of new sources like sensors on property and machines, connected devices, mobile devices, and digitalization of processes, together with customers' increasing willingness to share personal information. The consequences?

A maddening rush has begun as to who can gather the most data in an effort to achieve a competitive advantage. Just collecting vast troves of data is not enough. The data must be meaningfully processed to derive any value out of it. The success of using data to one's advantage therefore depends on the appropriate methodologies and tools to process it—and this is where AI comes in. By applying powerful algorithms, AI programs sift through huge amounts of data, generating novel insights and automating repetitive tasks. AI thereby offers great opportunities for optimizing existing and enabling new procedures, giving a competitive edge to businesses today and in the future.

To be able to leverage AI effectively, organizations need to combine a technological view with a business view—in other words, an understanding of what is technologically feasible today and how AI can create value for an organization by delivering business outcomes.

At a broad level, we can classify AI solutions into three areas—AI as UI (Ubiquitous Intelligence), AI as AAAI (Assisted, Augmented, and Autonomous Intelligence), and AI as II (Invisible Interface or no User Interface).

Artificial Intelligence as Ubiquitous Intelligence

You wake up in the morning and from your bed ask your digital assistant, "How is the traffic pattern on my usual office commuting route today?" It replies, "If you start at 8AM, your usual time, you will get to office in 45 minutes. There is a small congestion 5 kilometers from the office. I am actively monitoring and will give you alternative options once you start from home." The voice recognition, the natural language understanding of our question, the analysis of traffic patterns to get the right answer, and the translation of that answer into speech is all AI.

You get into your office and open your email. Your email is already organized into folders, such as In Focus, Social, Forums, Private, Junk, or whatever categories you have created. It is also tagged as important or not based on your organizational hierarchy (a mail from your boss is always important) or whatever tags you have provided to make it easier for you to prioritize your time management and actions you need to take. The classification of your email based on the To, From, Subject, and the Content, the natural language processing to extract the right keywords, the algorithm to determine what is spam or not, or who is important or not, is all AI.

You open your favorite news site and see that the news clippings are already personalized to your preferences. There are also few other recommendations on what is trending based on topics you follow. Curating news based on personal preferences and the recommendations for suggested articles involves using AI.

You open your favorite search engine, and as you type your query in the search box, the system suggests related words, as if it knows what you are going to type next. Then, as soon as you submit the query, the system not only lists the right content from billions of documents on the Internet, it also displays the most relevant ad that matches your query. The algorithm that magically reads your mind and suggests keywords to aid in your search, the page rank algorithm that computes the relevant pages to display, and the selection of the right ad using a real-time ad exchange are all AI.

These are just a few examples to highlight the fact that there is very little left in our day-to-day life that is not impacted by AI in some way. Everything we see around us is increasingly powered by an AI algorithm and while AI is gaining the status of ubiquity, we are sure that we have only scratched the surface regarding what AI can mean for us.

AI as AAAI—Assisted, Augmented, Autonomous Intelligence

Generally speaking, an intelligent system is something that processes information in order to do something purposeful. Whether you are searching for something on the Internet, shopping online, or seeking a medical diagnosis, intelligent systems are likely to play a key role in the process. These intelligent systems seek the best plan of action to accomplish their assigned goals (as assistive capabilities, as augmented capabilities, and as autonomous capabilities). You can think of these three roles of intelligent systems as a maturity continuum.

- *Assisted intelligence*: Primarily focused on improving what people and organizations are already doing in their day-to-day activities.

- *Augmented intelligence*: Gives an impetus to people and organizations as complementary capabilities so that they can do things they couldn't otherwise do on their own.

- *Autonomous intelligence*: Creates and deploys machines that are intelligent and adaptive by nature so that they can act on their own, completely eliminating human involvement.

Assisted Intelligence

Considered a weak form of AI, Assisted Intelligence optimizes the value of existing manual activity. Given clearly defined inputs and expected outputs, this form of AI automates repetitive tasks and provides recommendations for humans to validate and take action.

For example, assisted intelligence is already included in many cars—alerts about low tire pressure, not wearing seat belts, controlling speed, and activating the brakes if necessary. Although it will alert the user, the final decision usually does not belong to the vehicle.

Similarly, many administrative tasks like taking minutes at a meeting, organizing your mailbox, opening the page on your kindle when you last left off, etc. Are all done by some form of automated system. This type of automation, where the AI is assisting humans to do the same tasks faster or better, is a form of Assisted Intelligence. The humans are still in charge and make the final decision, but in the background the AI solution has done all the hard work to assist the human.

Let's take the example of a salesperson focused on closing a deal. She has to rely on many different systems to do her job. Email for communications with client and colleagues; SharePoint to find out customer information; Dropbox for documents; Skype, Slack, Yammer, Chatter, text messaging, and the phone for real-time communications; and business apps for order processing and customer relationship management.

On any given day, collating information from these disparate systems, drawing inferences, finding potential bottlenecks, and figuring out next steps, are daunting tasks. Now, consider an AI solution that extracts topics from each of these systems and assists the salesperson in evaluating what action should to be taken next. Should she contact the client to offer a discount, reach out to her solutions team to factor in additional value based offerings that client won't be able to turn down, or reach out to the internal competitive intelligence team to research a competitor's offering to develop a competitive value proposition?

This case of the salesperson is not an isolated one.

Beyond information assimilation, assisted intelligence solutions can also model complex realities of business processes, thus allowing the users to simulate business scenarios and weigh decisions before taking the final call. For example, a large construction company developed a simulation of their procurement related data—raw materials related cost, supplier's performance, enterprise wide practices the procurement officers adopt, the market behavior around supply and demand for raw materials, and the variations in those patterns for different city topologies. The model spells out more than 100,000 variations to consider and simulates the best option, thus assisting in huge cost savings for the company.

AI-based applications of this sort are also prime to assist in enterprise software platforms. For example, across the several ERP processes, if the corpus related to frequently faced problems and their effective resolutions can be harvested and fed into an AI program, then when a user is stuck in a process flow due to some inconsistency in data he is entering, the AI program can become active by offering recommendations to fix the problem, thereby smartly helping the user efficiently and timely finish the task.

Assisted intelligence solutions have a direct impact on several business performance metrics, such as employee productivity, revenues or margins per employee, and average time to completion for processes, to name a few.

How would you determine which areas or processes in your business are prime for assisted intelligence? You need to consider two questions: What products or services, if made automatically responsive, can improve user experience? Which of your current processes and systems would become more efficient with actionable insights?

Augmented Intelligence

The purpose of AI applications falling in this category is to lend new capabilities to human activity, enabling people and enterprises to do things they couldn't do before on their own. Unlike assisted intelligence, these AI applications fundamentally alter the nature of the task, thus creating possibilities of evolving business models or in some cases changing the business model entirely. The spectrum is broad and includes machine learning, natural language processing, image recognition, and neural networks.

In augmented intelligence, humans and machines are intertwined in a continuous learning and improvement loop. For example, routing optimization is a well-known logistics problem. By adding numerous other real-time data sources and feeds like weather patterns, local social events, political unrests/rallies, and traffic patterns to the existing static routing optimization mechanism, the AI solution can provide numerous alternatives, all extremely dynamic and intelligent enough to reroute if the driver chooses not to take the machine recommended top-ranking option. The human decision maker gets to see a high degree of granularity and specificity regarding the assumptions and evaluates the implications of accepting or not accepting machine generated recommendations. The machine tries to remodel itself by observing which recommendations the human decision maker accepts and which ones are rejected.

While the system learns a lot and models the ecosystem in a continuous self-learning mode, the humans learn a lot about the machine's behavior. Under these circumstances, the human and the machine share the decision rights.

In wealth management, augmented intelligence helps financial advisors better understand their clients and provide an enhanced portfolio determination beyond the usual aggressive, conservative, or balanced choices. The augmented intelligence system derives relevant signals from a holistic customer 360-degree view to give advisors insight into their clients' current and potential future financial needs. For example, information can be gathered about clients who may want advice on retirement planning or others who are celebrating the birth of a child. With this type of timely information, advisors can have important financial conversations with their clients at the right time, earning their trust for achieving financial goals.

The media and entertainment industry has been always in the lateral thinking and creative realm; however, there are a number of examples where augmented intelligence has changed the way the industry does things. Netflix movie recommendations are powered by an AI solution that learns about our personal preferences over time and gives suggestions depending on the choices we've made in the past.

How about AI attempting to become creative? That is precisely the partnership between IBM Research and 20th Century Fox recently set out to do, when they used machine learning techniques to produce the "first ever cognitive movie trailer." Watson analyzed hundreds of existing horror movie trailers to learn what kept viewers on edge. The AI solution then selected the 10 most engaging moments and a human editor created the trailer from those clips. The entire process took about 24 hours to complete, compared to the norm of month-long intensive manual efforts. Pretty cool, isn't it?[2]

In another instance, Natural Language Processing (NLP) capabilities were used to analyze thousands of movie plot summaries correlated to box office performance. The AI system became smart enough to recognize and plot patterns that would guarantee a successful box office performance. The movie "Impossible Things" was entirely created by AI by generating the initial premise and essential plot points, which the creative team then used as a point of reference to develop the final story.

The success of an augmented intelligence solution depends on whether it has enabled you to do new things that are either hard to do by yourself or that require out of the box thinking. Some of the potential proxies to use to assess opportunities are margin improvement, innovation cycles, customer experience, new business models to deliver non-linear revenue growth, and so on.

Augmented intelligence creates new possibilities. This aspect needs to be carefully managed in your enterprise, especially at the senior management level. Why? The new models will often challenge your senior executives to step out of stereotype thinking and explore alternatives and take calculated risks.

Autonomous Intelligence

In the continuum of the intelligent system maturity process, when a system gains enough self-learning capability and prediction accuracy that it can adapt to the changes in the environment without intervention, we call that state autonomous intelligence.

As such, the term autonomous intelligence is often applied to things that are technically robots but don't look like robots. Consider the example of a smart grocery cart in a supermarket that can navigate the isles and understands your shopping list to automatically pick up the items. Another example could be a home automation system where the windows autonomously adapt to light levels to optimize indoor environmental parameters, such as heating, cooling, lowering or opening of shades, or achieving the right level of indoor lighting.

---

[2]https://www.ibm.com/blogs/think/2016/08/cognitive-movie-trailer/

Autonomous intelligence also means handing over decision making to intelligent systems. The intelligent systems would work through enormous amounts of data, analyze the pros and cons, manage the associated processes, and then make a decision on your behalf. Off course, this demands a high level of prediction accuracy from the intelligent systems and at the same time humans need to trust the decision-making capability of the intelligent systems.

AI As Invisible Interface (No-User Interface)

Scott Jenson introduced the concept of a "technological tiller".[3] When we persist with an old design onto a new technology, thinking that it will work, we get into disastrous technological tiller situations. The term was inspired from the boat tiller, which was for a long time the main steering tool known to man. So, when the first cars were invented, rather than applying steering wheels as the steering tool, they used boat tillers. It was horribly difficult to control the car and often ended in serious crash situations. The adoption of cars with boat tillers was dismal. It was only after the steering wheel was added did the automobile become widely used.

There is a valuable lesson here: technological evolution demands radically different design approaches. When we ignore technological tiller, we usually end up with faulty products. However, when acknowledge technological tillers and bring in fresh design thinking, we achieve tremendous success.

Perhaps we are at the cusp of another technological tiller; this time it is about user interfaces. Given the advances in AI—conversational interfaces in the form of chatbots and commercial products like Siri increasingly playing a role in our personal life—is there a possibility of eliminating the user interface altogether?

We are surrounded by digital channels and devices screaming for our attention all the time: smart phones, tablets, laptops, smart-TVs, smart wearables, and portals notifying us about an event or prompting us to take action. The bulk of these requests and our responses take place through Graphical User Interfaces (GUIs). Of course, we have come a long way from character-based screens to elegant interfaces delivering rich user experiences. However, the biggest problem is the information overload and diminishing attention spans. Perhaps the situation is right for us to move away from designing GUIs, which requires the user's full attention, to designing calmer and highly intuitive human-computer interaction mechanisms to the core of the user experience. UI is not the product or the service; it is only a layer that allows us to access the product and services. So, the question is, if we can find a way to avoid the UI and still get to the product and services, then why not do it?

---

[3]https://medium.com/swlh/the-zero-ui-debate-e4b8bee4b742

Welcome to the world of no UIs.

GUIs facilitate three actions: present information, interpret user's instructions, and process the instructions. In this process, it's still our job as humans to understand the information, direct next actions, then evaluate outputs by comparing them with expected goals. The critical elements binding this process are context and intelligence. Is there another way of making this process simple? Let's take the example of you walking into your favorite bar and getting your favorite drink before even you settle down on the bar stool. The bartender knows you, knows exactly what drink you like, and knows that you just walked through the door. That's a lot of interaction, without any "interface."

In this example, the bartender is the UI. He understands what the user needs; he has all the context and all the intelligence and he delivers the perfect outcome. This kind of thinking is revolutionary in AI, if we can bring it to bear. On the one hand, this requires a lot more responsibility to create transparent experiences that tend to be highly relevant and contextual, all driven by intelligence (read as AI). On the other hand, this gives us incredible options for thinking out of the box and create open-ended experiences in which only important and viable information is presented to the user.

When we design products and services, we get carried away in the niceties and innovativeness of the features and we miss the critical aspect of knowing what is compulsory and what is optional—someone needs to do something. What are the minimum number of steps needed to allow them to do it? What is the minimum amount of information for each step? Minimally viable interaction is an interesting way of getting closer to an intuitive way of doing things, because it moves away from the focus on the product or service, to the *use* of the product or service.

For example, while designing an interface the designer sees it as a GUI, whereas the end user sees it as a means to book a flight ticket. Hence, it is important to put the spotlight on the end user. Ask what is the maximum amount of value the product or service can deliver to the end user in the minimum amount of time? What is the simplest possible way for the end user to get through the end-to-end process?

We've started talking to our machines—not with commands, menus, and frantic keystrokes, but by using human language. Natural language processing has seen incredible advances and we don't need to be a machine to talk to another machine.

No UI isn't really a new idea. If you've ever used an Amazon Echo, changed a channel by waving at a Microsoft Kinect, or set up a Nest thermostat, you've already used a machine that was inspired by no UI design thinking. We are beginning to move away from the touchscreen as the only way of interacting with machines to using more natural ways (haptics, gestures, computer vision, and voice recognition, all subfields of artificial intelligence) of interacting with the machines around us.

Until now machines forced us to come to them on their terms, speaking their language. The next step is for machines to finally understand us on our own terms, in our own natural words, behaviors, and gestures. That's what no UI is all about.

So far, our journey to design interfaces has been driven by the tool and the method, rather than what the final product or service is supposed to do. In the digital world, the act of doing becomes inseparable from the product or service. In other words, the interaction and the experience become the product and the service.

With conversational interfaces and voice and gesture-enabled interfaces gaining popularity, we are beginning to move away from "typing" into an interface to more and more "interaction" led interfaces. A look at the history of human-computer interactions will reveal that we have applied ever increasing amounts of computational and processing power to make the machines work harder so that machine response times are reduced. This approach has not simplified human machine interactions at all. Humans are hardwired to converse, so the most effective human-machine interface is one that is highly conversational in nature. We need to reimagine how we interact with machines and how they interact with us. We need to overhaul all our traditional interfaces to become simple, abstract, hidden, and ambient (the invisible interface).

The ultimate UI is no UI, all powered by AI.

Are New Advances in AI Worth the Hype?

Almost daily, we're hit with another news report or article glorifying AI use cases. Why exactly is the excitement so strong?

In a short span of time, we have seen a succession of major technological shifts (mobility, e-commerce, digitization, analytics, Big Data, IoT, automation, etc.), each one impacting the way we go about doing businesses and leading our life in general. When business executives and technology practitioners somehow come to terms with these technology evolutions, the AI wave and all the hype surrounding it poses a new challenge. What exactly is new about AI that may be relevant now? Why the attention to AI now? Is there anything new in AI worthy of the hype? Is this just "old wine in new bottles"?

Analytics, statistical analysis, optimization, regression, clustering, segmentation, etc., all existed long before attention magnified around the term "analytics." Airlines in particular have long used analytics for revenue management, pricing, and operational efficiencies. Marketing was using customer segmentation techniques long before analytics become the talk of the town. Fraud detection and credit scoring has been in action for several decades. Yet something was also new about the way analytics was done prior to the advent of Big Data technologies. Instead of following sampling approaches to do analytics, everybody started using Big Data technologies to run algorithms on the entire data set, thereby improving the performance and outputs of algorithms and sometimes stumbling upon insights that they were not able to find earlier.

Underappreciating the differences between the old periods and new would have been a mistake. We now have access to the processing power these technologies require. What could once be done in theory can now be done in practice and at scale. Prior incarnations of AI emphasized rules-based systems (also known as expert systems), formulated to automate reasoning. The new AI relies completely on learning patterns from data.

Business ecosystems have now become more digital. Previously, business applications and analytics systems required multiple integration points and data exchanges and often times humans had to manually connect the dots to carry out their tasks. But now with increasing digitization, links between applications and systems within organizations and externally are digitally native, allowing AI systems to interact directly with digital systems. These kinds of network effects accelerate AI adoption. As more and more machine-to-machine interactions happen, it opens up new opportunities for value creation (and destruction).

It is true that there is a lot of hype behind AI, but there is also an equal amount of reality. To overcome the confusion, managers need to embrace a mindset that allows them to listen to two perspectives simultaneously.

One perspective is about the potential that AI offers. Managers need to investigate AI technologies with relevance to their business so that they and their organization do not miss the opportunities. To assess AI's applicability to their business, they may pursue pilot projects to gain experience, even if the results are not Earth-shattering. They can also pick up certain existing business processes or applications to test AI approaches to find the strengths and weaknesses.

The other perspective is to question the promises of AI. On paper all these glorified examples may seem great. Managers need to truly understand and validate the relevance of AI for solving their specific problems and not get carried away by what the technology vendors say.

What Makes Adopting AI So Hard?

What problems you are trying to solve? Do you want to improve your internal processes like accounts, procurement, or HR functions? Are you looking to improve customer experience? Are there features in your product that would benefit from intelligence? Do you need your product to become more responsive to customers? Are there specific tasks that are repetitive, error-prone, and basically boring, which you can automate? Are there tasks where a little assistance can make your employees more efficient? AI technology can help you with all of these things and much more. If you have a chisel, you can use it to build any kind of statue. But to succeed, you just can't take the plunge and start applying the technology anywhere and everywhere. You need an AI strategy that aligns to the overarching business strategy. Whether you're improving your current business or building a new business model, AI should serve your business plan, not the other way around.

However, there are challenges!

AI can do magic, but before you can use it to solve problems, you need to look at it realistically and understand its inherent challenges. For example, AI has proved to be really good at classifying objects like tagging pictures, but it can't tell you a lot about what's in those pictures. In a business context, an AI application might be able to tell you whether users look delighted when they see a new product design, but they can't answer a bigger question, which is whether the product will be a success.

AI applications are data-hungry products. To train your AI application, you not only require a lot of data, but also robust data management practices. You'll need to identify data sources, build data pipelines, clean and prepare data, identify potential signals in your data, and measure your results. Organizations that are serious about AI have historically been proficient at acquiring and managing data as a strategic asset.

With the increase in the amount of data produced and analyzed within the enterprise and outside, there is a growing problem that enterprises are staring at—being drowned in a sea of data, analysis, dashboards, and data portals. Phrases such as "information overload" or "too many reports less insights" have become more common. Data-driven, a stepping stone towards AI, really means that all decisions and actions taken by the enterprise and employees are done by using the most factual and accurate data and there is a well-defined method of applying scientific analysis to this data to arrive at decisions best represents the actual scenario and context.

Achieving this strategic vision requires a significant change in the thought process of how problems are tackled. It is less about putting technology at work. It is about being conscious about how you are using (or not using) the tools (data and algorithms) at your disposal. You have to ask questions and not maintain the status quo; let's call it the "data-driven" strategy, which is a prerequisite to an effective AI strategy.

Prerequisite to AI Strategy: Data-Driven Strategy

Imagine adopting a strategy that provides notifications and recommendations to every employee irrespective of cadre and role, to carry out the tasks assigned to them optimally so that they not only achieve their individual goals but also contribute to the overall goal of the company. Imagine that these notifications and recommendations are provided to them at the right times through the right channels in a precise and prescribed way. The result?

Employees would become more productive and efficient, get more things done with less effort, get closer to customers, get more time to improve products or services, and in general elevate the enterprise to become a high-performing data-driven enterprise.

For the notifications and recommendations to become highly effective, they should be prescriptive in nature to cut down chances of ambiguity or dependencies and should have clear links to actions to be taken that result in measurable outcomes. This prescriptive intelligence, especially if it is based on the historical events-actions-outcomes paradigm, can result in dramatic improvement in the quality of work. It can also significantly influence and improve organizational productivity improvement metrics like first-time-right, on-time-completion, customer satisfaction index, employee-productivity as a percentage of revenue contribution, and many more. Is this possible?

Designing these notifications and prescriptive recommendations requires deeper understanding of processes, workflows, and dependencies between systems, and a broader view of the roles employees play in executing the tasks. The key to the effectiveness of the prescriptive model is that each recommendation should be highly contextualized to the task as well as to the employee who is supposed to carry out the task.

This is not a one-time effort. These notifications and recommendations need to be continuously monitored and measured for relevance and recall. Enough care should be also taken to calibrate the frequency of delivery of these options and recommendations. Overdelivery of notifications and recommendations will make employees build more dependency and expectations on these things. They will stop applying their judgment and in general won't learn. Similarly, underdelivery of notifications and recommendations will cause huge bottlenecks in the overall functioning of the enterprise.

What if business contexts change? Or if employees change roles?

This is precisely why constant monitoring of the notifications and recommendations generation mechanisms is required. At a macro level, these notifications and recommendations need to be reflective of changing business contexts and priorities, for example; new markets business is entering into or M&A opportunities business is pursuing or a new line of products it is launching, etc. At the micro level, an employee's responsibilities, job function, role, affiliations, and associations with specific groups or projects can change, causing both the notifications and recommendations to adapt to the changes.

How would you develop these notifications and recommendations?

Generating and delivering these notifications and recommendations can be achieved through several techniques: starting with if-then-else rules, heuristics, and embedding thresholds and alerts into every business process or through machine learning (ML) and artificial intelligence (AI).

AI and ML learn from the data and process more signals than humanly possible, leading to notifications and recommendations that otherwise could not have been possible or would not have been delivered in an adaptive and optimal manner. The advantage of AI and ML over other techniques is that once trained, they become self-learning. They observe the impact of the actions triggered by the notifications and recommendations; in short they are not hard-coded. They analyze the "actions" carried out by employees and their associated impact to improve the prescriptiveness quality of the notifications and recommendations.

Okay. How do you start?

The obvious option is to start small and target your most troublesome processes (need not be the most complex processes), target activities that are repetitive by nature but involve lots of hand-offs, and target tasks that require validation activities that today you do manually. Starting small not only makes the data-driven strategy more successful, but also provides a proving ground so that you can progress to more advanced notifications and recommendations such as those enabled through AI and ML. Over a period of time, progressively, these notifications and recommendations will mature, automated, prescriptive, and intelligent.

Staying Clear of Biases

AI models learn from data. By learning, we mean applying statistical and mathematical models to lots of known data so that you can arrive at a mathematical formula representing the behavior of that data set. When you apply the mathematical formula to run on a different set of data, one it has not seen before, it should correctly figure out the behavior of the unknown data set. It is the same as you experimenting on known things to find a pattern and then you apply the knowledge to unknown things to determine if your pattern matching is correct.

This learning process is vital to the success of your AI model. Many things can go wrong. If you get 100% accuracy on training data, chances are good that your model will perform terribly on real-world data. This is known as *overfitting*. On the other hand, if you don't decent accuracy on your training data, it will never be able to perform well on real-world data. This is known as *underfitting*. Too bad.

There are no hard and fast rules around where your model should lie in the spectrum of underfitting to overfitting. It is contextual to the problem you are solving. Some applications need a lot more accuracy than others; for example, you would not accept a self-driving car that is 99% accurate. Whereas you might be very happy with product recommendations that are accurate 60% of the time.

As AI becomes increasingly part and parcel of our lives, it is essential that we question how and why the AI recommended what it recommended. Most AI solutions are "black box," meaning the entire decision-making process is hidden (why did my GPS reroute me?). Hence, transparency and impartiality in AI is essential to building trust. Biased data and biased algorithms lead to biased predictions.

How exactly does bias impact the algorithms?

Data-Driven Bias

For any machine that learns from data, the output is heavily dependent on the quality of the data it used to learn. This is not a new realization. If the training data set itself is skewed, the result will be equally so.

Most recently, this kind of bias showed up in Nikon's face recognition AI solution. It started popping up alerts as "did somebody blink?" when taking pictures of smiling Asian faces. A clear case of AI solution learning from skewed example sets, certainly highlighting the problems that can arise when we do not pay attention to the biases in the data.

Bias Through Interaction

There are AI solutions like chatbots that learn through interaction. The idea is fairly simple. You interact with the machine and machine in turn learns from you to form the basis of subsequent responses. Microsoft's Tay, a Twitter-based chatbot, was designed to learn from its interactions with users. Unfortunately, Tay was exposed to a user community that used racist and misogynistic statements. As a consequence, within 24 hours Tay became a fairly aggressive racist and was forced to be taken down.

Emergent Bias

We all want products, services, and offers to be personalized for us. However, if we try to personalize too much, we tend to get into the trap of seeing things what we want to see, without contradictory or novel views. Let's take the example of our favorite news site. As we open, like, and share content, we keep feeding our existing belief set to the algorithm that does the personalization. The result is that we start living in a world of conformity belief.

These information biases have the potential to skew our thinking. A person who is only getting information that he likes from the people who think like him will never see contrasting points of view and will tend to ignore and deny alternatives.

Similarity Bias

Information systems that only provide results like "similar to" Tend to information asymmetry. Google News is designed to provide stories that match the user queries. This is explicitly what it is designed to do and it does the job very well. The result is a set of similar stories that tend to confirm and corroborate each other.

Having a different point of view supports innovation and creativity and in general acts as a powerful engine for improving decision making. However, if you are always provided "similar to" kind of information, you unknowingly are getting biased information.

Conflicting Goals Bias

Sometimes systems that are designed for very specific business purposes end up having biases that are real but completely unforeseen.

Maximize revenue and reduce operating cost are, for example, two metrics always in conflict with each other. If your AI solutions look at these two metrics in isolation and lead you to making changes, chances are that you may excel in both, but separately. However, when you start correlating your actions at a holistic level, you will realize that you have not really improved much. Rather you are in a vicious loop of prioritizing one versus the other.

Machine Bias Is Human Bias

Lastly, humans build algorithms (at least as of now), and as a result, algorithms end up reflecting our biases as well. Hence, before we start developing AI solutions, we need to develop an objective view of the problem we are trying to solve.

Are You Ready for AI?

Not every problem can be solved by AI (read as ML), and not every company has the wherewithal to apply AI at scale across the enterprise. So how do you assess whether your organization is ready to reap the benefits of AI?

To begin with, do you know what problems you want to solve? Do you know what you want to predict? Do you have enough data to build predictive models? Do you have the skilled people and technology platforms to build and train models? Do you already have statistical or physical models to give you a baseline for predictions?

There are three broad prerequisites to an AI approach: You must have lots of data, you must have capabilities to interpret the data, and you must have a way of making predictions from the data that help you achieve certain business outcomes. As you contemplate the introduction of AI into your organization, the next section outlines a list of best practices and critical success factors for your AI initiatives to succeed.

How to Define an AI Strategy

Given the hype that surrounds AI in general, business leaders need to critically determine the areas where AI can (and cannot) make significant material differences to the top or bottom line or even completely change their current business model. This is not an easy task and requires a dispassionate and objective view to address two specific areas: a) AI influenced industry disruptions that may significantly impact the current business model, and b) how to introduce and implement AI-driven solutions to usher significant changes to the way the company currently operates.

Identify Potential AI Use Cases

With every new technology buzzword, it is generally observed that companies go into a procrastination mode for quite some time. Reasons could be many—not enough appetite to experiment and learn, don't want to become the first few to take the plunge, uncertainty about the technology, or not enough confidence on the outcomes. The list can go on and on.

Hence, business leaders need to move quickly and identify opportunities where AI will create business value. A sure way to generate a list of business relevant AI use cases is to discuss the topic with everyone, including startups (you don't have to invent everything by yourself, rather look for collaboration and identify adjacencies or whitespaces that others are trying to figure out), do extensive research around the competitive landscape, and conduct design thinking sessions with your own business and IT leaders.

A very effective way of getting started is to brainstorm along two distinct themes.

Product/service innovation related questions:

- Does it provide a force-multiplier by improving efficiency or significantly reduce cost?

- Can it enable a better, efficient, and lower cost product substitution, thus altering the market dynamics?

- Does it integrate the enterprise value chain by optimizing or rationalizing processes, thus shortening time-to-market?

- Will it introduce features with hard entry barriers and thus enable you to charge premium and high margins?

Value realization related questions:

- Does it provide opportunities to leverage your current customer base and develop new services so that you can monetize your investments on data and algorithms?

- Is there any way to truly introduce differentiation and value-add services for non-linear growths by establishing network effects based on an ecosystem or a platform?

- Will it create uniqueness and establish strong entry barriers by preventing competitors from replicating your products, services, and offerings?

- Will it increase the cost significantly for your competitors to create alternative options, or will your competitors be more than willing to provide complementary offerings, thus adopting a sharing economy?

If the answers to these questions are more likely yes, cutting across business functions, corporate functions, and HR practices, you will likely end up formulating your own AI strategy. When you're doing this exercise, don't allow constraints to limit your answers.

Later in this chapter, we discuss a structured approach to defining AI use cases.

Assess Adoption Scenarios

Once you have identified your high-level AI use cases, start doing an assessment, which may further involve market studies, surveys with your existing customers, or even an independent validation by industry reputed consultants to determine what kind of adoption scenarios your AI initiatives will go through. Whether it will go mainstream or remain a niche offering. Essentially what you are looking for are early indicators—universal applicability, customer awareness and readiness, low switching barriers, low TCO, high ecosystem compatibility, scalability, and fewer regulatory barriers.

The more "yes's" you get against these indicators, the more you know that your AI solutions are ready for mass adoption and hence you should start formulating investment strategies and go-to-market strategies. If you are getting guarded answers against a majority of these indicators, you know that it falls into a niche offering category. Hence you need to start thinking about how to create a market for your AI solutions.

Assess Your Internal Capabilities

Having understood the adoption scenarios of your AI solutions, you need to turn inward and do a critical assessment of your own internal capabilities. Given AI's newness and inherent complexity, it is important to understand where the maturity of your internal processes, technology landscape, and employee skills stand. For any AI solution, the key ingredients are computing power, algorithms, data (lots of data), and skilled employees. If historically your company has not been at the cutting edge of technology, has not adopted agile development processes, has not focused on data-driven culture, has not encouraged innovation, then it is highly likely that even though you have stumbled into great path breaking ideas, you will struggle to bring those ideas into life.

Hence, it is important to work through the expectations for each AI initiative, prioritize the AI initiatives based on ROI and readiness to deliver, create a roadmap, and choose where to start.

How do you launch the AI transformation in your company? What building blocks do you need to establish and pay attention to on a continuous basis?

Launch the AI Transformation Program

We get carried away and pay a lot of attention to AI's transformative role in changing a company's core business model. However, to have the entire company marching on the AI transformation journey requires something else—a cultural shift in the company's way of working. The organization needs to become a learning company and not shy away from experimentation.

To become AI ready, in our view, an organization needs three broad-level calls for action.

Establish Sponsorship and Governance

From an enterprise AI readiness perspective, AI is less of a technology overhauling and more of a cultural shift, and both the business leadership and technology leadership have equal roles to play. Technology leadership needs to articulate how to make AI work to solve business problems and business leadership needs to envision how to adopt AI solutions to deliver business outcomes. These are not easy tasks!

To start with, you need to establish an organizational model consisting of an AI council that actively engages in the change-management process and sets up agile ways of working. You have to start with securing executive sponsorship and defining charters, roles and responsibilities, decision making protocols, escalation processes, and links to specific business objectives and processes. It is quite likely that some leaders in your organization will choose to wait and see (especially those skeptical about the AI implications). You need to engage more with those leaders to get some skin in the game, which means investing despite high levels of uncertainty.

Action Plans: Experiment, Fail, and Learn

AI solutions are all about rapid experimentation—fail fast and learn to improvise. Given the nature of AI solutions, it is foolish to adopt action plans based on traditional development methodologies. In the AI world, everything is a moving part. You need to design, develop, conduct tests, expose your working prototypes for user experience feedback, and test your deployment models on cloud working with your cloud partners and other ecosystem players—all at the same time.

Your plans should also align to the company's overall business strategy and should be prescriptive enough about how the senior leadership will adopt AI-influenced ways of thinking in daily business. It is especially important to be flexible and do course corrections to your plans as often as required. In the AI world, it is almost a regular occurrence that you start with one end objective in mind but end up creating many different possibilities.

Invest and Develop AI Capabilities

Considering the fact that AI is a new technology, you need to develop a clear plan of how to build needed AI capabilities and competencies, including an AI team of data scientists and data engineers, and the necessary infrastructure and algorithms.

You may consider starting with a centralized AI CoE to provide the initial thrust and support across the businesses. The AI CoE should not only focus on working closely with various business functions to create and enhance the AI use cases but also establish technology roadmaps, standards, and accelerators such as an AI platform, sandboxes for proof of concept development and pilot use cases, an algorithm marketplace to leverage and reuse algorithms built by others, a data catalogue to leverage and share data across the enterprise functions, and other best practices such as an AI playbook to evangelize cross-functional lessons across the enterprise.

The most critical task of the AI CoE is to attract and retain the right AI talent. AI talent is not unidimensional; it is a combination of different skills—creative thinking, willingness to research and go after unknowns, experimentation, business impact articulation, story-telling, technology prowess in data, algorithms, and the cloud. AI also brings with it a fair share of talent acquisition and talent management related challenges. Attracting the right talent, giving them the right environment to flourish, retaining them, and keeping their motivation level high is definitely not a trivial act. This is not a task to be left to managers and HR professionals alone; senior leaders of the organization need to stay involved, take ownership, and pay attention to the AI talent pool.

Given the scarcity of AI skills, it is understandable that you may not opt to set up in-house expertise to start with; you may collaborate with partners to set up innovation labs centrally or within your business functions to get them going along the innovation and experimentation journey.

How to Define AI Use Cases

Almost all AI use cases address three fundamental tenets: functionality, training data and algorithms, and an end-user need.

Using these three tenets, a use case can be described based on this formula:

Deliver (functionality) by learning from (data), thus fulfilling (user need)

For example, Spotify recommends new artists and songs (functionality) based on previously played genres/artists (data) in order to make relevant content easily and quickly accessible to users (need).

First and foremost, you must establish the *user need*. The user could be internal, such as employees, or an external user, such as customers. For example, a conversational interface for your employees to address various HR policy related queries fulfills a user need that is internal to the organization. If you are in the smart wearables manufacturing business, if you offer additional value-add services and recommendations around "staying healthy," Then you are fulfilling a user need that is entirely external.

A user need consists of several *functionalities*. The spectrum of possible functionalities can range from simple things like recommendations (assisted intelligence) to more complex things like automating decisions in your business process (augmented intelligence), to even more complex things like injecting appropriate advertisements, perfectly timed, customized to your personal preferences, and highly aligned to the content you are viewing (autonomous intelligence).

The following sections explain some of the most popular examples of artificial intelligence being used today. This is not an exhaustive list. In all of these examples you will find there are user needs met through certain functionalities using data and algorithms.

#1—Virtual Personal Assistants

They take many forms and leverage various technologies such as voice recognition, text analysis, NLP, and NLG. Some even make decisions for you such as automatically scheduling meetings based on incoming emails!

Alexa is a VPA made popular by Amazon. You use the wake word "Alexa" To start a conversation. As you start talking to Alexa, the machine will take your commands and start acting on it.

Amazon offers and Alexa Skills Kit to develop voice-first experiences with Alexa. From interactive games to smart-home integrations and drone control, anyone can design and build for voice. No coding knowledge is required. These voice-first AI solutions run in the cloud, meaning there is no installation required from the end user perspective.

#2—Video Games

Advancement in AI has greatly influenced how video games are played nowadays. Technology has moved from the crude forms of arcade games to non-player-characters (or NPCs) modeling real-world scenarios.

Consider the scenario that when an autonomous vehicle arrives at a stop sign, it must stop. Failing to do so could result in a human fatality. Part of the problem is that the image recognition algorithm must be able to recognize a stop sign. These signs can vary in their appearance due to low visibility, extreme weather conditions, paint pealing off, etc.

While developing the video game GTA V, the city of Los Angeles was extensively researched. Around 250,000 photographs and many hours of video footage were captured. These photographs and footage served as a training data set for the image recognition algorithm to identify and response to stop signs as if it were real life.

#3—Smart Vehicles

AI algorithms don't suffer from fatigue. This is precisely why integrating AI into long haul trucking is an attractive value proposition for many. Companies such as Google, Uber, Apple, Volkswagen, and Mercedes are heavily investing in self-driving automobiles powered by artificial intelligence.

#4—Consumer Analytics and Forecasting

Predictive shipping. In 2013, Amazon patented "predictive stocking." The idea behind this intelligent system is to reduce delivery times by predicting what consumers will want before they have even bought it

One example of the predictive shipping scenario: By observing historical buying patterns and determining the geographical locations based on ZIP codes, packages can be bundled together and shipped to a ZIP code-based warehouse or distribution center (even at the time of packaging and shipping one may not get the complete destination address). The AI in this case needs to exhibit a high degree of prediction accuracy so that there is no overstocking or understocking at the warehouse. Once the final destination is updated in the ordering system, the AI can then quickly get the last mile of delivery activities triggered.

Dynamic pricing. Retail needs to learn very fast to process all available data (internal and external), gain useful insights, and become agile at decision making. Price adjustments made for the store's inventory in seconds, in response to real-time demand, is much more effective than done manually with all the human mistakes. This is where AI comes into the play, by providing an option to optimize not only prices, but also business strategy, costs, and manager efficacy.

The number of variables that needs to be taken into account before setting price is very large: product seasonality, price elasticity, competitor prices, competitor markdowns and promotions, customer demand, retailer's desired margin, etc. It's nearly impossible to deal with thousands of products in real time without automated pricing algorithms.

An example of a rules-based algorithm:

> *If turnover KPIs are achieved then set the price equal to the competitor with the highest price. Otherwise, set the planned margin for the item.*

This rules-based algorithm may appear simple. Here are two examples where rules can become more complex and nested:

- Identify product clusters and correlated customer clusters with identical sales patterns to apply differentiated pricing rules.

- A/B smart price testing allowing a retailer to measure pricing rule influences on performance and forecast gross margins.

5—Finance

Fraud detection. Fraud detection has been rules-based for a long time. For example, a bank may create a rule that says something like this:

> *"If the customer is purchasing a product in a different time zone than normally observed and the cost is greater than $1,500, do not allow a straight-through transaction even if it is authenticated."*

Rules like this, if not continuously updated, can be gamed by adopting a brute force approach, where criminals can try different combinations of location, lower monetary values, and so on. Through supervised machine-learning techniques, the algorithm can learn from new patterns of fraudulent transactions, in real time, by observing every single transaction.

Credit decisions. To help banks make more informed credit decisions and become better at determining the lending risk, they need to look at attributes beyond the obvious ones used today. In addition, banks struggle with how to lend to people who don't have enough credit history. Banks need to look at external data sets like social network, job profiles, social styles, etc. To determine additional attributes. Today's consumer expects near instant decisions and AI can help drive them.

6—Chatbots

Advancements in NLP have propelled chatbots beyond just a Q&A system so that they understand the semantic orientation of each word in a sentence, derive true meaning to create context and relevance of what a customer is talking about, and then engage in a meaningful conversation.

Bank of America, one of the largest U.S. banks, uses a voice and text enabled chatbot called Erica to send customers notifications and help customers make better financial decisions.

JPMorgan Chase uses a bot called COIN to analyze complex legal contracts faster and more efficiently than a human can.

7—Social Networking

Facebook—With almost 2 billion active users on the platform sharing vast quantities of content consisting of text, images, and videos, Facebook has access to perhaps the largest data set on the planet. The largest data set is not the important; what Facebook does with it is the interesting part.

Facebook extensively uses various AI algorithms to make sense of this diverse data and provide insights to its users. For example, as soon as you upload a photograph, Facebook automatically highlights faces and suggests friends to tag.

Snapchat. Snapchat is another social networking platform that introduced "facial filters" To track facial movements and allow users to add digital masks that overlay their faces when moved. It uses AI technology, which was originally developed by a Ukrainian company called Looksery and uses machine learning to track facial movements in a video.

8—Real Estate

Matching buyers to new properties within minutes of being shared online—almost all of us have experienced this. However, there is more to it than simply matching keywords.

Roof.AI is using AI to automate tasks, generate leads, and integrate with Facebook messenger to revolutionize real estate activities.

Qvalue is using AI technologies to provide a completely different experience of house-hunting. You choose the style of home you love and instantly you get a list of the very best homes available, sale rank-ordered based on the qualities you love the most.

9—Drones

You're probably familiar with pilotless drones. They've been used by the military for years now. In recent years, drones have also made the switch from the military world to the civilian world.

Let's explore some examples of how drones are using artificial intelligence.

- *Life saving*: An engineer at the Technical University of Delft, one of the world's leading drone research hubs in the Netherlands, started to investigate if drones could reach a heart attack patient faster than an ambulance.

By working with ambulance services in Amsterdam, Alec Momont[4] established that the typical response time for a cardiac arrest call is approximately 10 minutes. Momont built a drone prototype that ships with a DIY defibrillator and is aiming to get there in six minutes. Momont's vision is for drones to be part of a wider emergency services response team and that someone witnessing a heart attack could call 112 (the equivalent of 911 in the U.S. or 999 in the UK) and the call handler would dispatch a drone. Using a two-way video connected to the drone, a medic could talk the witness through the necessary steps of using the defibrillator.

One can see the obvious advantage to having such technology in rural areas or difficult-to-reach locations.

- *Hover camera*: Lily Camera[5] hovers in the air and is powered by artificial intelligence. Weighing only 238 grams, the self-flying camera can be carried around. It's like having your own self-flying personal photographer. Once in the air, the drone automatically finds and follows you (its owner), while recording your everyday life from a completely new angle. It's all possible thanks to advanced artificial intelligence facial recognition algorithms.

  At the time of writing this book, even though Lily camera was a much-hyped and innovative AI offering, it has gone through rough weather.

- *Logistics*: In December, 2016, a British farmer named Richard Barnes received an order placed on Amazon for a bag of popcorn and an Amazon Fire TV Stick.

What was different about this delivery? It only took 13 minutes for the goods to be delivered from the time of order and was fulfilled by using an autonomous drone!

10—Text Analytics and NLP

Natural Language Processing (NLP) helps a machine make sense of human language. NLP allows humans to communicate with the machine in natural language. Let's explore some examples of text analytics and NLP.

Customer Reviews and Sentiment Analysis: Consumers often leave comments or reviews on specific products or services that they purchase. This sort of unstructured data requires AI technologies such as sentiment analysis and POS Tagging (Part of Speech) to make some sense out of it so that business can find out how consumers feel about their product, brand, or service.

---

[4]http://www.alecmomont.com/projects/dronesforgood/
[5]https://www.wired.com/story/the-drone-company-that-fell-to-earth/

AdTech: Have you ever looked at products in Amazon and then moments later noticed similar products being displayed in your Facebook or Twitter feed?

By tracking what you like and what you've viewed and the comments you post and share, machine learning can, with reasonable accuracy, place marketing creatives in your newsfeed on social channels, thereby improving conversion rates for business.

AdTech is such a lucrative area that companies such as Twitter have launched developer initiatives to encourage the development of AI-based software to drive online sales.

The most important ingredient necessary to define an AI use case is *data and algorithms*. Without vast amounts of high-quantity data, it is impossible to develop a killer AI solution that solves real-world problems. The performance and usefulness of an AI solution is heavily dependent on the quantity and quality of the data. AI use cases require careful considerations of the data sources. (This includes enterprise business data, partner ecosystem related data, external data such as social networks, weather conditions, publicly available data, open government data, census data, and other third-party curated data sources.)

Although not an exhaustive list of data sources, the following list is a high-level view of different kinds of data required to solve AI specific problems:

- Call center recordings and chat logs can be mined for content and data relationships as well as for answers to questions.

- Streaming sensor data can be correlated with historical maintenance records, and search logs can be mined for use cases and user problems.

- Customer account data and purchase history can be processed to look for similarities in buyers and predict responses to offers.

- Email response metrics can be processed with text content of offers to find buyer segments.

- Product catalogues and data sheets are sources of attributes and attribute values.

- Public references can be used for procedures, tool lists, and product associations.

- YouTube video content audio tracks can be converted to text and mined for product associations.

- User website behaviors can be correlated with offers and dynamic content.

- Sentiment analysis, user-generated content, social graph data, and other external data sources can all be mined and recombined to yield knowledge and user-intent signals.

AI will no doubt continue to impact every aspect of our personal and professional lives. Much of this will occur in subtle ways, such as improved usability of applications and ease of finding information. These may not necessarily give an appearance of powered by AI, but they are. Over time, as AI matures in the continuum, we will get into a scenario where AI will take the form of "applied human knowledge".

Hence, no matter what business you are in and what role you play in your organization, you need to develop strong foundations for AI.

Conclusion

Can you afford not to get on the AI train? You can't. But don't jump on the wagon blindly; map out where you're headed before taking the plunge. The notion that artificial intelligence is a magic ingredient that makes all things better will lead to costly mistakes. Understand what you're doing, why you're doing it, and the limitations you face. Both the limitations of AI itself and the limitations of your organization.

It would be very, very helpful to know what the future holds for artificial intelligence in business. Unfortunately, it is also very hard to predict. Why? Whenever we are faced with a prospect or uncertainty, we tend to look back into history and develop our mental map by extrapolating from there to the future. This approach for AI may not work well.

A Sloan review article[6] outlines some very interesting perspectives. In three months, Joshua Browder developed an AI application to help people win disputes over parking tickets. In the first 21 months, the AI application registered a 64% success rate.

Our natural reaction to this might be any of the following. Put in another three months and the AI application might deliver 100% success rate. Or if you are happy with the success rate, you want to expand the applicability and take the AI application to other locations. Or if this thing works for parking tickets, then we can seamlessly extend the AI application to solve other problems like utility bills disputes!

Although appealing, the reality is that extrapolating the future success and applicability of AI from one field to another field is not that straightforward. Why is this particularly difficult for AI? To get to the answer, first we need to familiarize ourselves with three principles.

---

[6]https://sloanreview.mit.edu/article/on-the-road-to-ai-dont-ask-are-we-there-yet/

The Pareto Principle

Common sense tells us to invest efforts where the benefits are greater. The Pareto Principle says that 80% of the effects come from 20% of the causes. By focusing on a few scenarios (the 20%), AI can deliver significant values that address the majority of effects (~80%).

This is exactly what Joshua focused on. He found out that appeals courts dismissed most parking tickets for 12 main reasons. His AI application focused first on these 12 reasons and achieved a fair degree of success. Now, if we want the same AI application to look at more and more reasons, each new reason incorporated will require more and newer data sources, more training time to analyze rarer cases, and consequently may result in a diminishing contribution to the total number of successful appeals.

The takeaway: You need to define the goals or outcomes for your AI initiative with a reality check. By putting in more effort or expanding the scope, you may achieve incremental benefits, but it will come at increased cost and increased complexity.

The Ninety-Ninety Rule

Tom Cargill of Bell Labs found an interesting pattern in the software development process, "The first 90% of the development effort accounts for the first 10% of the project timeline. The remaining 10% of the development effort accounts for the remaining 90% of the project timeline." The underlying problem is that the developers first attempt to get the simple and medium complex features completed, which in most cases constitutes 90% of the scope. The remaining 10% of the scope is all about complex or highly complex features, which typically take much more development effort.

AI solves problems by learning from data, no complex coding of rules. While this approach eased out the development effort, it created a larger dependency on the data related areas, mainly how exhaustive is your data sources so that your AI application can cover almost all the scenarios possible. In reality, even though we claim to have access to abundance of data, much of the data is full of noise and less of signals. Thus, while your best and concerted effort may result in solving 90% of frequently occurring scenarios, the effort required to solve the rarely occurring events will be much more harder. In the case of the parking ticket application, much less data is available to train the AI application to address the "long tail" of dismissal reasons.

The takeaway: For AI in business, efforts required to solve an expanded scope may increase disproportionately compared to the earlier scope.

Hofstadter's Law

Hofstadter's Law says that time estimates for how long anything will take always fall short of the actual time required—even when the time allotment is increased to compensate for the human tendency to underestimate it.

The pervasiveness of AI and its popularity as a term has led to rising expectations as a panacea to all sorts of problems, and therein lies the problem. First are the expectations around quick results, second are the expectations around tolerance.

Given a clearly defined set of legal rules, Joshua was able to figure out how his AI application is going to solve the problem. Having this clarity upfront also meant that there was a certain amount of certainty around time, budget, and outcomes. In short, the problem statement and the expectations were clearly aligned. Not all problem statements follow the similar alignment. For instance, when Joshua extended his AI application to a more complex scenario (issues related to refugee asylum in UK), it took longer for him to develop the AI application.

The takeaway: AI solutions do not follow traditional project management guidelines. You need to continuously experiment to achieve the desired results. Hence, learning fast, and continuously weighing cost vs. benefit and tolerance for error are the keys to developing AI applications.

In this chapter, we covered several fundamental aspects of AI, including the maturity continuum of AI. The next chapter discusses how to reimagine a competitive advantage in the AI world, including the role of AI in the boardroom.

References

1. https://hbr.org/2016/11/the-simple-economics-of-machine-intelligence

2. https://medium.com/swlh/how-to-create-a-successful-artificial-intelligence-strategy-44705c588e62

3. https://www.strategy-business.com/article/A-Strategists-Guide-to-Artificial-Intelligence?gko=0abb5

4. https://www.fastcodesign.com/3048139/what-is-zero-ui-and-why-is-it-crucial-to-the-future-of-design

5. https://www.interaction-design.org/
   literature/article/no-ui-how-to-build-
   transparent-interaction

6. https://techcrunch.com/2016/08/15/using-
   artificial-intelligence-to-create-invisible-ui/

7. https://techcrunch.com/2015/11/11/no-ui-is-
   the-new-ui/

8. http://usblogs.pwc.com/emerging-technology/
   ai-everywhere-nowhere/

9. https://www.mckinsey.com/industries/
   automotive-and-assembly/our-insights/
   how-advanced-industrial-companies-should-
   approach-artificial-intelligence-strategy

10. https://www.edge.org/conversation/
    pattie_maes-intelligence-augmentation

11. http://variety.com/2017/digital/news/
    best-artificial-intelligence-applications-
    media-1202624021/

12. http://www.mediaentertainmentinfo.com/2017/
    09/top-10-areas-artificial-intelligence-is-
    leading-automation-in-media-industry.html/

13. https://www.salesforce.com/blog/2017/11/why-
    ai-drives-better-business-decision-making.html

14. https://sloanreview.mit.edu/article/on-the-
    road-to-ai-dont-ask-are-we-there-yet/

Reimagining Competitive Advantage in the AI World

AI as an area of research and fascination has been around for a long time. It's a field that once disappointed its proponents in terms of progress and is now at the forefront of solving real-world problems. It's rapidly entering into our daily lives as it expands into activities commonly performed by humans, often raising concerns about our very existence.

As AI is beginning to realize its potential in achieving human-like capabilities, businesses are asking how they can harness AI to take advantage of the specific strengths of human and machine.

Sensing, thinking, and acting have been key to human evolution, but then, when you say machines can do these things as well, it raises quite a lot of eyebrows. Humans tend to exhibit heightened capabilities in general intelligence (we call it *fast parallel processing and pattern recognition*), whereas they tend to slow down when it comes to logical reasoning (we call this *sequential processing*), owing

© Soumendra Mohanty, Sachin Vyas 2018
S. Mohanty and S. Vyas, *How to Compete in the Age of Artificial Intelligence*,
https://doi.org/10.1007/978-1-4842-3808-0_2

to careful evaluation of implications/rewards. On the other hand, computers/ machines are becoming good at pattern recognition in narrow fields (called *narrow intelligence*), yet they are superfast at logical reasoning provided there are clearly defined inputs and expected outputs.

Based on the characteristics it exhibits and the contours of interaction boundaries as well as its limitations, AI can be placed into four categories, discussed in the following sections.

The Sense-and-Respond Machines

The most basic AI systems are purely reactive; they are designed to sense an event and programmed to provide a pre-determined response. These basic AI systems neither have the ability to understand what led to the event, nor do they have the ability to form memories or use past experiences to make informed decisions. Deep Blue (IBM's chess-playing supercomputer), which beat international grandmaster Garry Kasparov in the late 1990s, is the perfect example of this type of machine.

Deep Blue was designed to sense the pieces on a chess board and determine what move its opponent is playing. Based on the opponent's move, it was programmed to respond with the most optimal move from among the various possibilities. However, apart from having access to a huge library of chess moves and rules, it didn't have any concept of the past, nor any memory of what moves were made before. It was designed and programmed to sense the present moment and respond with an appropriate move.

Similarly, Google's AlphaGo (based on neural network techniques), which has beaten top human Go experts, uses a much more sophisticated analysis method than Deep Blue's, but it still can't evaluate all potential future moves. By applying neural networks and sophisticated algorithms, these machines have progressively become better at playing specific games and beating human experts at their own games, but they aren't adaptive to other situations easily, whereas humans are. In other words, these machines can't function beyond the specific tasks they're trained to work on and can be easily fooled.

They can't adapt to changing scenarios (if you change the rules of the game, they will stop working and need to be reprogrammed). Instead, these machines will behave exactly the same way every time they encounter the same situation. Actually, this behavior of being consistent is their strength in a different context. It brings about a heightened trust quotient—consider that you would want your autonomous car to be a reliable driver all the time, every time.

Limited Memory Machines

This type of machine has limited memory and limited reasoning capabilities. Self-driving cars do some of this already. For example, they observe other cars' speed, proximity, and traffic signals. These observations are added to the self-driving cars' reference library to simulate real-world representations and act based on what is pre-programmed. But this library of information about the real-world representations are only transient, it lacks the ability to learn from the environment in a continuous manner, the way human drivers compile experience over years behind the wheel. For self-driving cars to achieve human-level cognition, they need continuous updates to their pre-programmed representation of the world.

So how can we build AI systems that are capable of full representations of the world, remember the experiences, improvise to adapt to the changing environmental conditions, and learn how to handle new unseen situations?

Theory of Mind

This is the point where machines start to truly exhibit thinking and reasoning capabilities on their own. Machines in this more advanced class not only have the capability to form representations about the world, but also have the capability to interact and learn from other machines. In psychology, this is called "theory of mind"—thoughts are continuously evolving and reasoning capabilities are deciding trade-offs.

The theory of mind is central to human evolution, because it allowed us to learn from social interactions. Assuming someday machines will achieve the same thinking and reasoning capabilities as humans, they need also to be aware of human motives and intentions; otherwise, the human-machine coexistence will become at best difficult, and at worst impossible.

Self-Aware Machines

The pinnacle of AI development is to ultimately build systems that can not only think and reason but also exhibit consciousness.

This is, in a sense, an extension of the "theory of mind" possessed by machines that are self-aware. For example, understanding and interpreting "I want that item" is a very different statement from "I wish I had that item". It's all about being conscious and being able to predict other people's feelings. For example we may assume that aggressive drivers are either in a hurry or are impatient, so we let them go. Our self-driving cars need to exhibit the same understanding and give way. Without a theory of mind, we can't make these kind of inferences.

While we are probably far from creating machines that are self-aware, it is obvious that we will need quantum leaps in processing power for machines to exhibit a combination of vastly different types of sensing, thinking, and problem-solving capabilities—the hallmark of human intelligence. For example, today's much hyped self-driving car doesn't exhibit what we would consider common sense, such as pulling aside to assist a child who has fallen off her bicycle.

It begs the question: How can businesses leverage AI to gain a competitive advantage?

AI in the Boardroom

The very idea of artificial intelligence (AI) taking a seat in the boardroom may seem preposterous and far-fetched. After all, experience, lateral thinking, judgment, shrewdness, business acumen, and an uncanny ability to spot trends ahead of time are the critical skills required for the kinds of complicated matters that boards often deal with. However, at the same time, it is also true that AI is addressing some extremely nuanced, complicated, and important decision making processes. So, AI taking a seat in the boardroom may not be a bad idea at all!

Businesses are going through heightened sense of volatility and there are strong headwinds in the form of technology initiated disruptions. We all know that the cost of bad decisions is high, and if these bad decisions are made at the board level then the implications are further magnified. While there are numerous examples of successful new product launches, mergers and acquisitions, and digital transformations, at the same time there is an even larger number of instances where many of these decisions have turned into failures, dragging the company into the ground. A report from Innosight (a growth strategy consulting firm) shows an alarming trend— "The 33-year average tenure of companies on the S&P 500 in 1964 has narrowed to 24 years by 2016 and is forecasted to shrink to just 12 years by 2027".

Over the past five years alone, companies that have been displaced from the S&P 500 list include many iconic corporations such as Yahoo, DuPont, Staples, Dell, EMC, and Safeway, to name a few. While the reasons are many and some were beyond the control of corporate leaders, it is true that many of these disruptions were driven by a complex combination of technology shifts and economic shocks.

The signs were there, but these companies missed the opportunities to adapt or change. For example, they continued to apply old business models to new markets (similar to the tiller effect discussed in Chapter 1), were slow to respond to disruptive competitors in low-profit segments, or failed to adequately envision and invest in new growth areas that often takes a decade or longer to pay off. In the mean time, the explosion of "decacorn" companies across industry sectors like transportation (Uber and Lyft), financial services (ANT Financial and SoFi), aerospace (Space-X), real estate (WeWork), healthcare (Outcome Health), and energy (Bloom Energy), as well as everything in the technology space has been accelerating the disruption by introducing new products, business models, and services to new markets and customers.

In short, businesses have become too complex and there are blind spots— most leaders are under the false notion that future competition is coming from existing players, whereas the reality is that new competitors are entering, serving your customers in an entirely new way. You need to respond appropriately to these new set of rivals and for that your board and CEO need to make good decisions with the support of intelligent systems.

Leaders across every industry are failing to understand these disruptive forces and need serious help in developing some kind of methodical approach to the madness. The Innosight paper recommends five essentials:

1. *Spend time at the periphery:* A good place to start is to continuously scan the horizon and understand what the emerging unicorns are doing. Your corporate strategy and research teams need to spend time online or in physical spaces experiencing new products and services and reflect back on the impacts to your core business model. By doing so, you can spot early warning signs and opportunities that could cause massive shifts in value.

2. *Focus on changing customer behaviors:* You can no longer be under stable assumptions; you need to become customer obsessed, observe your customer habits minutely, and seriously evaluate their changing behavior patterns, no matter how tiny these patterns appear to be. While you may think you are providing a large bouquet of services, in reality you might find there are new entrants solving essential jobs at cheaper rates and thereby eating away at your business.

3. *Adopt a future-back strategy:* The conventional approach to strategy starts by analyzing what has happened his- torically, assessing the implications of future trends, and arriving at a SWOT kind of analysis. Then you extrapo- late from today to formulate the strategy for tomorrow. While this approach has worked reasonably well in the past, given current business conditions where VUCA (vol- atility, uncertainty, complexity, and ambiguity) is the new normal, this approach constrains strategic choices and

can prevent a company from objectively assessing their current model. In today's dynamic business conditions, leaders are overwhelmed and worried about not living up to their responsibilities.

To stay future ready, organizations need to adopt a different strategy. Instead of starting from the present and looking forward, they have to envision a future environment and business portfolios based around changing customer needs. The vision of the future only serves as the starting point. You then need to move backward in time to develop a set of strategic initiatives and innovation interventions for the present. The underlying assumption is that tomorrow may not resemble today.

4. *Embrace dual approach*: Transforming the core business is only going to give companies incremental linear growth. Hence, companies need to adopt a dual approach like the digital platform companies have shown. Discover new growth opportunities outside the core, and then invest and govern them separately.

5. *Assess the cost of inaction*: All said and done, it requires a strong conviction to prioritize and marshal your resources including capital and talent to embrace the dual approach. When it comes to decision making, leaders tend to take the safe way out—they push new growth ideas to the back burner. While there may be valid and perfectly acceptable reasons to do so, you need to continuously validate your strategy given the pace of technology evolution and multiple forces of disruption at play. There is no longer the notion of annual corporate strategy and blueprints. Nowadays you need to have a quarterly corporate strategy and blueprints and estimate a price tag for inaction by measuring the impact of lost opportunities.

The question is, how do you prepare for such dynamic and fast changing scenarios?

One plausible solution is to incorporate AI supplements into the practice of corporate governance and strategy. This is not about automating leadership and governance, but rather about augmenting intelligence using AI, for both strategic decision-making and operations decision-making. How do you do this?

AI could be used to improve strategic decision-making by analyzing capital allocations and highlighting concerns—maybe your company is cutting spending on research and development while your competitors are increasing investment in the R&D and M&A areas, thereby actively funding startup ecosystems and becoming better at market intelligence to identify potential new entrants moving into your market space. Similarly, AI can be used to improve operational decision-making by analyzing internal processes and systems and determining employee productivity, predicting churn, highlighting inefficiencies, etc.

Being a new technology, AI has its fair share of naysayers—executives are uncertain of its business case and believe that human capabilities are critically important to capture the returns from AI in the enterprise.

Having lots of data and producing insights is not the key. How you consume the analytical results is the real game-changer. In the digital era, no doubt organizations have matured in the data and analytics areas. By investing in infrastructure, tools, processes, and talent, they were able to improve their analytical insights production capabilities more quickly than they were able to improve the consumption abilities. As a result, despite the fact that their analytics production capabilities were improving, the analytics consumption gap is widening—the opposite of what you would hope and expect.

Managing AI technology demands new leadership skills, including those required to implement, govern, and make right use of the analytical insights.

Using AI to Create a Strategic Advantage

"Datafication" of the world and rapid evolution in prediction technology have gotten us to solve well-defined but complex problems like recommending movies, diagnosing cancer, and creating autonomous vehicles, to name a few. However, just using the technology, no matter how advanced, does not deliver a competitive advantage. For technology to advance the business strategy, it must be embedded into business processes, customer interaction touch points, partner ecosystems, and into the day-to-day life of the enterprise workforce.

No doubt, technology has gotten better and smarter. As a result, AI has forayed into tasks that require creativity and intelligence, such as creating movie trailers, composing music, performing facial recognition, and detecting emotions. They have also gotten better at conversing with humans, including interpreting and extracting insights from images, machine generated data, text, and unstructured data. Given such rapid progress, all in the last few years, there are good reasons to get excited that technology might deliver a "AI strategist" app that can directly assist CEOs in their strategic decisions.

How do we build a "AI strategist"? It can't be developed by itself, at least for now. It requires significant human involvement to begin with. We human beings are unique in many ways; we can think outside the immediate scope of a task or problem. AI lacks this capability. After a lot of training, it can flawlessly execute a well-defined task, but if you change the problem definition, it will falter.

In other words, for AI to exhibit human-like thinking, it needs to act in concert with humans and learn how to seamlessly execute conceptual and analytical operations—including problem definition, signal processing, pattern recognition, abstraction and conceptualization, analysis, and prediction. Of course, this is not to say that AI is incapable of learning these higher-order skills—but that state is far away, too far away.

For AI to demonstrate effective business strategy development, it must demonstrate *reframing skills and continuous learning capabilities*—the process of redefining and reanalyzing the problem with feedback.

Amazon provides an excellent example of how AI and humans, in a continuous learning mode, execute the business strategy. The company has several AI systems (supply chain optimization, inventory forecasting system, sales forecasting system, profit optimization system, recommendation engine, and many others). These systems are intertwined with one another and with human strategists to create an integrated, well-oiled AI ecosystem. For example, if the sales forecasting system detects that the popularity of an item is increasing, it triggers a series of communications: update the inventory forecast, which triggers an optimization process in the supply chain system to update inventory across warehouses. This causes the recommendation engine to push the item more, the profit optimization system adjusts pricing, and the marketing system and launches real-time campaigns and discounts. The resulting impact in turn feeds back to the sales forecast system. These are only some of the first-order effects; there are numerous further interactions between the systems and partner ecosystems that occur downstream. While many of the operations happen automatically, human beings also play an important role: they design experiments and review data traces to make the AI systems continuously learn and improve. Humans also pay specific attention to extract higher-order insights from anomalies and patterns captured by AI, that essentially serves as food for thought and as the impetus for Amazon's next strategic moves. For example, if the item gaining popularity is not usual, then it requires a certain amount of investigation. Is it a long tail item but gaining popularity because of certain uniqueness that was not observed earlier? Is it because the item suddenly found a fan following due to another related but costly item falling out of favor? How seasonal or short-lived is this moment?

How can businesses create an effective AI strategist app? At a broad level, there are three requirements, described next.

A Future-Back Goal

To start with, you must define the desired outcome. Since strategy can encompass anything and everything, it is important to have a well-defined area to begin with; for example, you could focus on competitive intelligence with a special focus on unicorns. Human strategists must provide the initial set of questions and help resolve ambiguity. The AI app can then learn to improvise based on continuous feedback and eventually become better at what it is supposed to do.

The opposite of not having a specific goal is to go on a wild goose chase. There is always this danger with powerful technology—it leads us to become preoccupied in thinking about what it *can* do rather than what it *should* do. We then invariably get caught up and get driven by AI's capabilities, rather than addressing the problems that we need to solve. In other words, if all you have is a hammer, then everything looks like a nail.

A Human-in-the-Loop Approach

The problem statement, data, algorithms, conversational interfaces, feedback mechanisms, automation capabilities and most importantly the human strategist (either to intervene or to refine) must form a tightly integrated continuum. This integration is critical, because the human and the technology should have only one objective—to optimize for the global outcome rather than for individual tasks. Why? Let's take the problem statement as, "evaluate a new business opportunity". This will need deeper research around competitive threats, strategic fit to the organization's vision, evaluation of whitespaces, build vs. buy decision prudence, etc. All these areas are related but are also a sub-domain on their own. If they are analyzed separately, the AI strategist will be no closer to an answer unless there is a mechanism to integrate the findings and the human is involved as an integral component to resolve trade-offs, thereby generating new insights.

As the AI strategist takes on increasingly complex questions, human beings, instead of running around to collect data (which is where they spend a lot of time today), with their unique ability to understand the broad context and connect insights from disparate spheres, will spend more time making sense of the data and feeding those lessons back to the AI strategist, which eventually will help make the AI strategist perform better.

At this rate, it is not absurd to think that one day the AI strategist will become smart enough to surpass the human strategist. It will be an interesting day for sure.

A Well-Designed No-UI

The AI strategist must be able to communicate its findings and recommendations to people; conversely, people should be able to understand, examine, and validate those recommendations and provide feedback to the AI strategist.

Humans are able to think outside the box and carry out higher-order tasks of reasoning, validating, and reframing because we communicate effectively. If the AI strategist operates as a black box and creates outputs that are incomprehensible, it loses its value. People cannot interpret the outputs and therefore cannot build deeper and richer insights through successive reframing. To avoid this situation right at the onset, the AI strategist must be designed to be transparent and conversational.

How Important Are Cost, Quality, and Time?

AI as a discipline has progressed rapidly. It has emerged out of the labs and is already making dramatic changes to our day to day life. Some AI applications have even reached human-level performance in many aspects of vision, conversational speech, and problem-solving. As a result, every industry is actively considering AI initiatives within its enterprise. However, since we are in the midst of such unprecedented disruptions, there is no methodology or prior use cases one can follow. Important questions include: How much does it cost? How do you define ROI? How do you measure the quality? How much time should it take? There are many more related questions, including what will happen to our jobs!

No doubt, with today's robust AI technologies, we are transforming the way we have been working. In many sectors, we have automated activities that are repetitive and mostly back-office operational tasks. Interestingly, AI has also started to lend its intelligence to fields that require less repetitive manual labor and once seemed immune to automation, such as law, education, journalism, etc. In short, AI has started augmenting human minds, not just muscles.

In the midst of all these happenings, it is important to acknowledge that there's no shortage of work that can be done only by humans (refer to Chapter 1— human judgment becomes invaluable). The challenge that organizations need to address is not a "world without human involvement" but a "world with rapidly changing human labor-related skills".

Certainly costs and quality parameters are important for any business to stay competitive and meet customer expectations, but the AI development lifecycle needs to be looked at differently compared to other business application development lifecycles. The initial AI development cost may appear to be expensive, but once it's developed and the AI system starts running in the environment, it can scale to accommodate enormous workloads without a fuss.

The other important advantage is that AI systems continuously learn from data as well as from the human-interactions, thus making the ongoing run and maintenance costs negligible, if not zero. For example, once you have developed an AI system to analyze the lab reports and report findings, you can scale it to do the same analysis for millions of lab reports at minimal incremental cost. This aspect of AI paves the way for organizations to replace or increase productivity of expensive knowledge workers as AI automates many tasks. The result? An opportunity to reexamine currently expensive bottlenecks and see how AI can overcome them.

With machines, we can expect execution discipline, precision, and consistent results. Your AI radiologist will deliver the same output every time, all the time. With this consistency in place, organizations can incrementally refine and improve quality over time.

While reducing costs and improving quality are important, AI also addresses a fundamental business goal—faster time to market. For example, the current loan approval processes typically take 10-30 days. Getting a loan, even a pre-approval, doesn't happen overnight. With data increasingly available to support all the information that goes into a loan processing application, AI can dramatically reduce the time required to process the loan. If you find a house to your liking in the morning, there is no reason why your loan application will need a month to be processed. Ideally you should be getting the keys to that house by the afternoon. Loan approving officers (if they continue to exist) will simply need to be able to swipe your credit card to close the transaction.

Lawsuits are at the other end of the extreme. They can take considerable amount of time to resolve. Some of the time is in gathering data and preparing the case, some is in delays due to court congestion, yet more is due to deliberation and settlement. This entire process, with multiple bottlenecks and dependencies, could be shortened by applying an AI solution that can research and prepare the case along with references to past judgments and then provide an early indication of how much time it would take to arrive at the outcome. Certainly, it will help people from just going round and round in circles, wasting time and money, if they are provided with enough factual information about where their case stands!

Medical diagnoses are another example of cumbersome processes. Many times the delay happens because of non-availability of past medical history. Secondly, even if you have the past medical history, you are dependent on availability of the right physician in your insurance covered healthcare provider network. What if you need a simple diagnosis to understand how bad your condition is? AI can shorten the lag time considerably by analyzing your past medical history and matching your current symptoms with a vast library of diagnosis results. The potential for faster preliminary diagnosis can certainly help hospitals and clinics and at the same time improve health outcomes through earlier detection of conditions that require quick medical attention.

As AI's usefulness begins to outweigh its uncertainties, what do organizations need to watch out for?

Too Fast to Manage

We are fortunate that the slow pace of our current processes allows us time for monitoring and management. If (when!) something doesn't work as expected, we usually have time to perform root cause analysis, find any bottlenecks, instigate fixes, or perform course corrections. However, when machines run our processes, we may not have time to step in and manage the situation, because everything will be running at a machine pace. We may not get the opportunity to intervene before the money is gone from a risky loan, the guilty walk away, or the patient suffers from an incorrect treatment. In the race to optimal speed, the machine pace (extreme automation) may lead to risky outcomes.

Too Fast to React

Besides many other reasons like improved productivity and improved responsiveness, one of the key reasons for expediency is to stay ahead of your competitors. With AI you can create stiff entry barriers for your competition. However, it won't likely work out that way. Just like you are seriously thinking about AI, your competitors are as well. The result? When everybody wants to stay one-up, it creates a huge amount of instability in the ecosystem. You no longer have the luxury of market leadership or competitive advantage; it suddenly has become a level playing field.

Too Fast to Learn

AI solutions are data hungry. The more data there is and the more varied it is, the better the AI outcomes. In a bid to outweigh competition, if every enterprise seriously acquires as much data as they can, suddenly you end up in a situation where there is no longer any differentiation due to data. The other issue is that the value of data may decay quickly. The rate of decay may exceed the rate of availability. The Netflix prize is a classic example: the company offered a prize for a substantial improvement in its algorithm that recommends movies based on historical customer viewing behavior (DVD-by-mail rentals). After awarding the prize, Netflix found that the algorithm was not useful on the new data (video streaming behavior). When AI depends on data, the rapid decay of the data's value may actually render the algorithm less useful.

Are AI Learning Scenarios Unpredictable Enough?

What if a collision happens between an AI operated vehicle and a human operated vehicle? Quite naturally, the sensational headlines would play to our biased thinking that the machine is to blame. This scenario is not fiction. It actually happened in Las Vegas, where a minor collision happened between a self-driving shuttle and a human operated truck when the delivery truck backed into the front bumper of the shuttle. The AI operated vehicle was trained to stop if there is an obstruction on its path, and it behaved exactly as it was supposed to behave.

This incident highlights a crucial gap in AI and human interactions. AI systems are typically trained in context where there are no external factors with adversary objectives. The focus is to learn from observations and, while designing such systems, our first intent is to get something working.

Consider how the incident between the AI operated vehicle and the human operated vehicle unfolded. The AI operated shuttle accurately recognized the obstruction, predicted a potential collision scenario, and stopped. This is the strength of AI—observing many inputs and processing those quickly to arrive at a reasonably accurate decision.

Since the collision happened anyway, the question is: what else could the AI operated vehicle have done to avoid the collision? This is where it becomes difficult. It could have honked loudly, which might seem like common sense and fairly a risk-free option to humans. It could have swerved away from the approaching truck, but this seems more difficult and riskier than a honk. For imperfect AI (still learning and improving) faced with uncertainty, a reasonable decision is to stop and do nothing.

The core of the problem is information asymmetry—perfect information versus imperfect information. When we are thinking about human-machine interactions from a game-theory perspective, access to information (or lack of information) changes the outcomes radically. If the AI system and the human don't know what the other will do—both have imperfect information—then the situation is extremely volatile and unpredictable. If AI has perfect information, that is, it knows what the human will do, then the uncertainty no longer exists (at least for AI with perfect information). The situation plays to the AI's strengths. Similarly, if humans know what the AI will do, but the AI systems have imperfect information, we are creating a scenario that plays to the AI's weaknesses.

Let's take the example of job applicants. Once the job applicants figured out that AI systems were looking at an absence of certain keywords to filter out job applications, they got creative and included every possible keyword that would show up as an important and critical skill in their resume. Awareness of how the algorithm worked meant applicants could manipulate the system to their benefit.

There are heightened concerns that the preoccupation with autonomous driving and creating human-appearance like robots is distracting us from the more enterprise relevant, and potentially more transformational, applications of artificial intelligence in business. While there is nothing wrong with solving problems that we had never imagined we would be able to solve, prudence must be applied by business leaders. We must not stay stupefied by these fascinating applications of AI, but learn from these examples and apply novel ideas to solve business problems.

How can managers can bring sanity into the AI operated world? Let's take some inspiration from a scenario that played out during the period of the Cold War.

Lieutenant Colonel Stanislav Petrov noticed that the Soviet information systems were sounding alerts about incoming nuclear missiles from the United States. He was responsible to make a well-informed decision and he had to act fast: Should he authorize a retaliatory attack or hold on and make further investigations? Fortunately for all of us, Petrov chose to investigate. He realized that a real attack was unlikely because of several other factors—one of which was the small number of "missiles" reported by the system. After his death in May 2017, a report credited him with "quietly saving the world" by not authorizing a retaliatory attack.

The real reason behind the false alarm was due to system's inability to accurately distinguish the light signatures of sun's reflection off clouds to light signatures from missiles once they become airborne.

Businesses face similar (although hopefully less consequential) questions about whether and when to remove humans from their decision-making processes. There are no simple answers. As the false alarm incident demonstrates, a human can add value by carefully evaluating a system's recommendations and its implications before taking action. We can surely question the efficacy and robustness of the AI algorithms. Humans develop the algorithms, hence humans could add more value by helping the classification system prevent misclassification.

If our world was full of autonomous machines, when and why would they want our assistance? In the nuclear attack scenario, the machines, while correctly predicting the imminent attack should also highlight the consequences of the desired action—worldwide destruction. Perhaps we should build enough checks and balances to weigh in the consequences of a certain action, even if it is the right thing to do.

With immature AI, the machines should assess and recognize their own state of inaccuracy and initiate human intervention automatically. For example, if there are insufficient observations, humans are more likely to utilize their breadth of experience, confer with other humans, and learn from other instances in ways that machines cannot (yet). Classification does not have to be a binary state (missile or no missile); AI systems, when lacking confidence, should request human help.

Fortunately, business decision-making often differs considerably from the fully automated machine world. The pace of digital business may have accelerated, yet many business decisions can still afford time for a second opinion where humans can play the role to confirm or deny. In business situations where time does not allow additional steps of human intervention, the AI systems should keep asking for additional training to correct errors.

The key question is: Since AI is still maturing, should we take a wait and watch stance or should we take few pragmatic steps, clearly knowing that there are implications?

Managing Immature AI

Despite the advancements in AI technology and even as organizations continue to push the boundaries of what's possible with AI, there is a wide gap between the promise and reality of AI.

While at one end, we have many successful implementations of AI solutions, at the other end, we do have many unsuccessful attempts. This is largely because AI is still a young, immature field.

Algorithms are competing with, and often winning against, expert humans in complex games such as chess and Go. But at the same time, simple customer service chatbots are failing to address the real pain points with customers and at times can be more annoying than helpful.

AI has the potential to apply data and algorithms to offer quick decision making; however in many scenarios it seems the algorithms are themselves biased, thus not giving an impartial view on the outcomes. Facing this contradiction, it can become difficult for executives to get a clear view about the usefulness of AI, and hence there is the temptation to adopt a wait and watch policy.

So, how are executives pragmatically incorporating AI into current business processes?

Clearly, AI can take the information you have and generate information you don't have. From this perspective, you can expect AI's primary job to enhance a knowledge worker's ability to make use of the predictions (there is a high probability of someone defaulting on a payment), whereas the primary job of the knowledge worker is to apply judgment (seems like an anomaly, as this person has never defaulted) to improve AI's prediction performance.

For quite some time, the set of recognized prediction problems were statistical and operations research related, such as inventory management and demand forecasting. However, over the last few years, image recognition, driving, asset maintenance, talent acquisition, health care, and many other tasks that were solely the purview of humans have also been framed as prediction problems.

As more and more tasks that were solely under the purview of human-prediction work get reframed as machine-prediction problems, organizations need to shift from training their employees in prediction-related skills to judgment-related skills. In short, organizations need to focus on creating an ecosystem and processes to enable teams of judgment-focused humans and prediction-focused AI agents working in concert to deliver business outcomes.

The effectiveness of AI technologies will be only as good as the data they have access to, and the most valuable data may exist beyond the boundaries of one's own organization. Organizations will need to look at their current inward looking data strategy and open up to develop strategic alliances that depend on data sharing.

For example, BMW, Daimler, and Volkswagen, even if they compete with each other, formed an alliance to buy a Berlin-based digital mapping company (HERE) to create a real-time platform that would track and monitor driving conditions (traffic congestion, estimated commute times, accidents, or vehicle break-downs on routes, emergency services, public utilities, and weather patterns). These are based on data collected from cars and trucks from each brand. Alone, it would have been prohibitively costly for one company to build such a platform, but together they created a sufficiently robust data platform that provides improved customer service and creates new revenue models, such as subscription based fees for municipalities, private emergency service providers, citizen services companies, insurance companies, and many more.

AI will affect different organizations in different ways, and many of the changes will have direct implications to managers, such as how they plan work allocations, how they review the work in progress, how they provide interventions, how they manage a combination of AI and employees, and so on. Additionally, these changes bring considerable risks as well.

- *Replacement risks*—AI promises to provide assistive intelligence to human performance in a number of ways, which is good as it improves human productivity, but at the same time AI also redefines certain activities, thereby completely eliminating human involvement. The reality is that human roles and positions are not immune any more. Insurance companies have already started using AI (instead of human agents) to offer customers insurance plans. As the push for smart machines, smart factories, and smart plants continues to progress at a rapid pace, the manual labor force, whose workday is largely intertwined with operating machines, will find their jobs are taken away by AI agents.

- *Dependence risks*—In the long run, as AI becomes completely embedded into business processes and in a sense runs the enterprise functions, a strong dependency will emerge. Unless enough care and strong governance models are in place, this dependency will create new vulnerabilities, creating deadlock situations between AI agents and human-in-the-loop and leading to potentially ineffective operations. Errors and biases may creep undetected into algorithms over time.

- *Security risks*—Today, we manage security through surveillance, monitoring, and access controls. Our current security management practices are designed to detect unauthorized activities by humans. We are beginning to realize the implications of security threats from software. Imagine when enterprises are fully running on AI agents and use sophisticated algorithms that deal not only with mission-critical applications but also with sensitive data. The risk of AI stealing information from other AIs will be very real.

- *Privacy risks*—As AI takes on the role of information worker in an enterprise, the AI itself will become a source of information over time—collecting, analyzing, and governing fairly sensitive corporate and customer data. Widespread use of AI will certainly raise many data privacy and ownership issues (we are already seeing stricter data management and protection regulations in the form of GDPR for the EU). The black box nature of AI systems will pose additional challenges for human-machine collaboration scenarios. Unless you know why the AI system did what it did, you really won't be able to trust the AI system. This is where organizations need to come out with clear ethical standards and draw the line between AI autonomy and human ownership.

How will AI impact and disrupt the way enterprises do business? And how should executives plan for the upcoming decade of disruption?

Innovative organizations reinvent themselves; they demonstrate that they are able to fail fast and learn faster. If you are graduating with an AI technology degree from a prestigious school or you have invested your time wisely to get trained on the latest AI technologies, would you pursue a startup that is focused on a niche AI-based product offering, or join a company that wants to build innovative AI applications, or use your academic skills to collaborate with scientists in other fields to conduct advanced research?

The opportunities presented by the first two options are glamorous (thanks to the press) and highly rewarding, consequently driving the salaries of AI skilled resources through the roof. Why is there such a rush to hire AI talent? The answer lies in the fact that companies are sold on the idea that AI will magically create new business segments or will help widen the gap between them and their competitors.

In our view, the third option of advanced research is equally important for long-term sustainable and ethical use of AI in all spheres of life and society. We keep hearing optimistic views that AI (and other digital trends) will create new industries and new job categories that will outshine whatever job losses it causes. There are two open questions regarding these views: How will these new industries be created? How soon will they come?

New industries are not created overnight; it takes decades to transform pathbreaking ideas into viable commercial value propositions, which eventually leads to setting up new industries and creating new jobs. Given this long journey, the real question is, will these new industries and new job categories emerge fast enough to maintain sufficiently high employment levels in the economy and offset the jobless impact due to automation and AI?

Companies lapping up AI talent are primarily focused on creating AI solutions to automate and optimize, not to create new industries. There are numerous examples already proving this mentioned point: Uber, Lyft, Airbnb, Amazon, and many other algorithm-powered companies are either focused on improving efficiency out of current business models or creating new business models by removing process bottlenecks (intermediaries). These two methods, while introducing step-change efficiencies to businesses as well as end-users like you and me, are not necessarily creating new industries or new jobs.

Hence, it is equally important for companies and governments to find a way to use the already scarce AI talent to contribute to the discovery and pursuit of scientific and technological advances that could lead to creation of new industries and new types of jobs.

Skills to Succeed: EQ Skills for Evolving AI Economy

Baron-Cohen popularized the empathizing-systemizing theory. People bring different skills and strengths to do their jobs effectively. Empathizing skills come to aid at identifying, understanding, and resolving conflicts, motivating teams, and responding to the mental states of others, whereas systemizing skills come to aid at analyzing, understanding, reasoning, troubleshooting, planning, and predicting outcomes. According to Baron-Cohen, women score higher on empathizing and men score higher on systemizing. This doesn't mean that women can't excel at systemizing or men can't excel at empathizing.

Continuing our conversation on the impact of AI on job roles, the roles that are likely to disappear over the next decade are not limited to a particular industry or cadre in the corporate hierarchy—the impact is far and wide and across all levels. AI agents will replace not only truck drivers and assembly line workers, but also radiologists, financial planners, insurance agents, and security guards, to name a few—all traditionally male-dominated roles. It is also abundantly clear that as more and more work becomes automated, there will be a significant shift in demand for the skills like judgment, empathy, compassion, influence, and engagement (skills in which women more often excel). For simplicity, let's call these as emotional quotient (EQ) skills.

We are heading toward a world where AI would act as a great leveler. How? Take the example of radiology. The AI agent may determine that your radiology scans indicate cancer, and you will be at ease when a human sits down with you and explains the best course of action. Imagine, instead of a human, the machine tells you in a boring and monotonic voice what your treatment plan and next steps should be! Similarly in case of business, the AI agent may suggest what operational improvement actions are required in the company to deliver better margins. It is still much more effective for a human to lead the charge to persuade people to execute the recommendations. Imagine the machine sending out prescriptive actions to your employees!

In the AI economy, where machines will increasingly do the "systemizing" type of work (where men excel over women), there will be a greater need for "empathizing" type of skills (where women excel over men). Usually we put a premium on our technology skills and often we tend to downplay our EQ skills. However, given the AI led future knocking at the immediate horizon, all of us—men, women, and organizations—need to start paying attention to the importance of the EQ skills.

How can you get started developing your EQ skills? Just like any technical skill, a person's EQ skills can be improved through training and by providing them with the right support systems. Here are three steps to get started:

- *Create your own EQ baseline:* Humans are social beings. We work in teams; we communicate with each other and together we solve problems. However, when it comes to providing feedback, most of us are very hesitant to criticize someone's interpersonal skills because such feedback pushes us into uncomfortable situations and we become defensive. Many sharp, effective people have no idea that they need to improve on their EQ skills because they simply haven't paid attention to the subtle indicators from their peers and teams. Hence, the first step is to become self-aware of your own EQ quotient and pay attention to feedback you've been given, especially comments along the lines of, "You are difficult to work with," "You are too argumentative," "You need to do a better job of reading others body language."

- *Acknowledge that EQ skills are important too:* We have always paid way too much attention to systemizing skills, whether it is systems, technology, engineering, mathematics, or related fields. Although every role has an EQ angle to it, historically we have pushed it to the back burner. For example, doctors are well trained to identify and treat disease. Even though they are expected to sit down with patients and personalize the treatment plan to suit the patient's preference and lifestyle, they tend to ignore or delegate this specific aspect of relating to someone else. You must first be clear about the outcome you are delivering, not the activity—is getting the diagnosis right the most important measure of success? Or is it actually improving the patient's health? If the latter is true, then it is equally important for you to start focusing on your EQ skills.

- *Reframe learning and knowledge management as EQ management:* Learning and knowledge management functions have always taken a myopic view of business needs. The focus is to train you for tasks to be done not for roles of the future where more than your technical skills your interpersonal skills will become much more important. You need to honestly assess your EQ quotient. You need to find a coach who will give you honest feedback and mentor you, and you need to consciously work on the improvement areas. The difficult part of EQ is that we are hesitant to agree that we have EQ gaps. None of us want to admit our EQ needs work, and we have this notion firmly printed on our minds that our EQ is inborn and unchangeable. We are wrong on both accounts.

As AI becomes increasingly embedded into everything we do in our professional and personal spheres, the "softer" aspects of our skills need to become much more important.

Use Case: AI and Amazon Flywheel

More than any data-driven company, Amazon has delivered stupendous services and products with its audacious vision for an AI-powered enterprise. At the center is Amazon's "flywheel," which ensures various parts of its enterprise functions work in tandem, by feeding into each other to deliver business outcomes. Another interesting concept that propels Amazon to keep delivering amazing innovative products and offerings is the "future-back-to-the-present" approach. At Amazon, any initiative starts with the end in mind—a "six pager" pitch starting with a speculative press release describing the finished product or service.

AI is at the core of this flywheel, where algorithm powered innovations in one part of the company are extensively leveraged to accelerate the efforts of other teams. Amazon meticulously examines every opportunity to monetize what they build. For example, during the early days of Amazon, they built robust data management and AI platforms to efficiently run their own business. Later they realized if they can offer these internal capabilities as a paid service to outsiders, they would not only make money, but they will also get an opportunity to understand how people are using the AI services and what they are doing with it. That way they could churn out more meaningful offerings in the future.

Amazon had its humble beginning as a bookseller. First it established a marketplace and launched a powerful e-commerce platform. Perhaps Amazon was the first company to truly realize the importance of a platform. It opened up its marketplace to other retailers, thus establishing a sharing economy that took advantage of its e-commerce platform. It built warehouses to fulfill orders for customers, and then offered fulfillment services by Amazon to its network of marketplace business partners, logistics, and distribution companies and other channels. Amazon found a way to monetize the excess computing capacity (AWS) they had built to support the business during the busiest shopping seasons.

"Amazon Go" leverages the computer vision and AI algorithms to offer a cashier-less shopping experience to supermarket shoppers. The intent is not to open thousands of Go stores across the country, but to offer this AI-powered retail infrastructure to shopkeepers as a subscription. Another monetization idea in the making.

Right in the formative years Amazon realized the potential of AI and started investing in it. In the early days, Amazon's AI talent was scattered across divisions and cross-leverage of the innovations and ideas were sporadic. The top-down push to institutionalize AI across the company brought in significant changes. The company put "serving its customers" as the key driver for anything they would do, which resulted in islands of AI innovations to collaborate across projects and share their solutions with other groups, thus putting the flywheel in motion.

The Conversational Interface Effect

Amazon's Echo line of products powered by the voice platform Alexa also sprang from a six-pager, delivered to Bezos in 2011 during an annual planning process called Operational Plan One. The vision was to come out with a low-cost computer with all its brain in the cloud that you could interact with over voice—you speak to it and ask questions, and it engages in a conversation with you and provides answers.

Building Echo required a level of AI prowess that the company did not have at that point. However, thanks to its approach of working backward from an imagined final product, the high-level blueprints included features that were critical to making Echo. The voice recognition feature in particular demanded a level of conversational AI that was yet to be invented. For example, activating the AI system by using a "wake word" ("Hey Alexa!"), hearing and interpreting natural language commands, filtering out noise and delivering non-absurd answers, etc. Were not conceptualized anywhere.

In addition, for Echo to be commercially successful, it has to be cheap, extremely effective in voice recognition capabilities, and be able to deliver never before experienced end-user satisfaction. For all this to happen, Amazon had to build an AI system that could understand and respond to natural language queries in noisy conditions. The biggest bottleneck was where Amazon would get massive amounts of conversational data to train the AI.

The answer to this question was found in the Amazon flywheel. It started leveraging AWS (its cloud infrastructure offering) to create a speech recognition service (Alexa) that became a valuable asset beyond its original scope of fulfilling the Echo's mission. Once they developed Echo as a far-field speech recognition device, they saw the opportunity to do something bigger. First, they integrated Alexa into their other products and offerings: you converse with Alexa to access Amazon Music, Prime Video, your personalized recommendations from the Amazon shopping website, and other services. Second, they expanded the scope of Alexa to become a voice service by allowing developers to create their own voice-enabled applications (known as "skills") to run on Echo.

In the beginning Amazon struggled to find conversational data to train Alexa, but once customers began using Echo, Amazon steadily gathered real conversational data. This conversational data became a powerful learning and experimental test bed—once you have a device in the market, you get access to real data that is so fundamental to improving everything, especially the underlying platform.

Democratizing AI

Amazon realized that if they could externalize what they were doing internally using their cloud platform and their own data scientists, they could generate tremendous value. They started focusing on how to simplify AI. Another epic six-pager was in the making.

While the tactical aspect was to add machine-learning services to AWS, the broader aspect was a grand vision. How could AWS become the destination choice to build AI applications for all and sundry? You don't have to invest on platforms, you don't have to have data scientist teams of your own—you just need to define the problem you are trying to solve. The rest is all on AWS.

In 2016, AWS released new machine-learning services—Polly (a text-to-speech component), Lex (a natural language processing engine), Rekognition (a deep learning offering for face recognition)—and launched a more comprehensive AI platform called SageMaker. These offerings allow AWS customers (spanning from giants like Pinterest and Netflix to tiny startups) to build their own AI-powered applications.

AI Way-of-Working

Amazon democratized AI within the company and various units in a decentralized way. They created a specialist central group to define standards, architectures, and best practices. This group's primary objective is to evangelize and promote machine learning across the enterprise in a highly collaborative fashion. Several examples highlight how different units within the company leveraged lessons from other units to solve problems.

- The fulfillment team wanted to better predict which of the eight possible box sizes it should use to package a customer order. They leveraged the algorithms developed by other groups and applied them to solve their own problem.

- Amazon Fresh, the company's grocery delivery service, needed a better way to assess the quality of fruits and vegetables. They leveraged new algorithms developed by their Berlin-based team using IoT sensors in the delivery vans to determine the freshness of fruits and vegetables in transit.

- Amazon Go leveraged a new AWS service called Kinesis to process streaming data from hundreds of cameras to track the shopping activities of customers in the store.

- Amazon's Prime Air drone-delivery service, still in the prototype phase, built a new AI solution to enable its autonomous drones, drawing on knowledge from the rest of the company and figuring out what tools to use.

A New Business Model

Shopping using Amazon has become a part of our life. We visit the website, shop for items, place them in the "basket," and pay for them, and then Amazon ships them to us. Right now, Amazon's business model is shopping-then-shipping.

We all have noticed Amazon's recommendations when we shop; the basic idea is to increase the basket size. The recommendation engine offers suggestions of items that you may find interesting to buy. It is nothing but a prediction system using "product-user similarity" algorithm. Considering that there is a match making happening at scale (millions of items to match with millions of customers' buying preferences), the prediction system does a reasonably good job by accurately predicting what we want to buy about 5% of the time. In other words, we actually purchase about one out of every 20 items it recommends. Not bad!

Amazon continuously collects information about us. In addition to our searching and purchasing behavior on their website, it knows not only what we buy, but also what is our preferred time to do online shopping, where we are located, how we pay, how many times we browse through before we finally make a decision to pay, how many times we have returned products, and many more. The prediction system uses all these data points to continuously refine its algorithms. The goal is to keep improving the metric from browsing to actually buying.

What happens when the prediction system's accuracy level comes significantly closer to 100%?

Will it become more profitable for Amazon to ship you the products it knows you will want to buy rather than wait for you to order them? A change in the business model from shopping-to-shipping to shipping-to-shopping!

Because the prediction system has now reached the threshold of being 100% accurate, Amazon can ship you products it knows you will want. You don't have to order anymore. You decide in the comfort and convenience of your own home which items you want to keep and which ones you want to return. This new approach offers two benefits to Amazon. First, since Amazon proactively delivers products to your door step, it indirectly persuades you to not to go elsewhere to do your shopping. Second, by delivering products to your doorstep, it influences you to swipe your card, which otherwise you were considering to do for some time but might not have gotten around to. In both cases, Amazon benefits by gaining a higher share-of-wallet. Turning the prediction accuracy far enough creates a new and profitable business model for Amazon.

What are some of the challenges of this new business model?

In the current shopping-to-shipping model, Amazon picks up the products we want to return. However, in the new business model of shipping-to-shopping, there will be many more returns than observed today. So, Amazon will have to invest in infrastructure (perhaps scale up their fleet of trucks or expand partnerships with local logistics companies) to manage product delivery and returns process efficiently.

You can ask, what if they were to launch this new business model at the current prediction accuracy (that is, 5%)? By launching sooner, Amazon will gain a lot in terms of access to newer data and will improve its prediction systems accuracy much faster. In addition, it will create a new entry barrier for competitors. The new business model will attract more shoppers, more shoppers will generate more data to train the AI, more data will lead to better predictions, and the cycle continues.

What do we learn from the Amazon flywheel example? First, when AI is embedded into enterprise functions and processes, it improves the prediction capabilities and in turn has a significant impact on the business strategy. Second, there are increasing returns to investments on AI; however the timing of adopting AI as a business imperative matters a lot. Adopting too early could be highly experimental and costly, but adopting too late could be fatal.

Companies face two questions in light of all of this. First, they must develop a better understanding of how fast and how far they can embed prediction capabilities in their businesses and applications. Second, they must focus on developing a strategy to continuously assess and gain business benefit of applying prediction capabilities to their business.

Putting AI to Work

AI has emerged out of the labs and entered the world of mainstream business. AI as a service offered on the cloud is simple enough for all to adopt; however, there is a lot of hard work that goes into managing the interplay of data, processes, and technologies, and solving problems that are relevant for your industry. If you are under the impression that you can simply subscribe to AI platforms on the cloud, throw all your data at the pre-built algorithms, and you will magically produce "intelligence," then you are completely wrong.

The following sections discuss AI examples across industries and corporate functions, to give you a sense of how AI is used to solve problems.

Marketing and Sales

Gone are the days of mass marketing campaigns. In this increasingly digital world the best way to draw attention is to personalize. AI enables us to offer personalized services, advertising, and interactions. We have already seen the impact of successful personalization in the retail industry. Using not only the data shared by customers but also a large variety of data that is available in the web, retailers, for example, have developed loyalty apps that can reside on our smart phones. Based on our location, time of day, and proximity to supermalls or retail outlets, the loyalty app can push hyper-personalized offers to use in real time.

In sales and marketing, AI can provide augmented intelligence capabilities correlating millions of individual data points with information on general consumer trends and then build a real-time marketing system that will potentially deliver thousands of customized offers in real time. This is a complex task that marketers today would probably take a week to deliver.

"Next best offers" are positioned at the intersection of a customer's needs and the company's catalog of offerings. The key is to understand what the customer really wants (explicitly said and unsaid), do a match with the company's offerings, and close the deal to the satisfaction of the customer as well as the company. In case of the insurance industry, this would mean building a model that reflects the needs of customers as they pass through various life stages. The model will leverage complex algorithms that can crunch more than 1,000 static and dynamic variables consisting of demographics, prior policies, prior agent interactions, economic conditions, segmentation of other customers, thus matching this customer and what products they have taken. The result? A potential to significantly increase cross-selling. The insurer can also use the model to reflect on how its agents are performing, thereby designing training programs to upskill agents who are not up to the mark.

These examples demonstrate the importance of a rich supply of contextual and specific customer data. If you have a well-defined problem statement and data, you can generally validate a proof of concept within four to six weeks and then put together a detailed plan for the data infrastructure and the resources required for a full rollout.

Research and Development

R&D is a fascinating field and the problems in this field are generally complex. They require deep domain and technical expertise and are often driven by many experimentation cycles. AI solutions in the R&D area can expedite the experimentation cycles, often predicting the outcomes by simulating millions of data points.

For example, in industrial manufacturing, AI can correlate real-time sensor data from machines with past maintenance data and OEM benchmark data to predict when the machine is going to fail.

The problem in the R&D space is lack of availability of data. Companies wanting to better their product designs need to digitize machine operations data. Consumer electronics sector will need to focus a lot more on 3D design data and user experience data. The pharmaceutical and healthcare sector will need to focus on clinical trials data and drugs efficacy data. Given the knowledge and expertise required to develop AI solutions purpose-built for R&D, companies must engage domain experts to systematically curate data and help improve the prediction capabilities.

Operations

Operational practices and processes have standard operating procedures, generate a wealth of data, and have measurable outputs against each task. Quite naturally, operational tasks are well suited for AI supplements. Rapidly evolving technology is transforming every field, and manufacturing is no exception. Cost pressures are rising as industries become increasingly competitive. Heavy industries like chemical, power, and thermal, with longer gestation times and long-term ROI, are facing the brunt, not only from within but also due to the fast-moving external factors. For example, labor that was once readily and cheaply available is now hard to get.

Manufacturing trends in itself are changing; made-to-order is the new way. Just-in-time, lean manufacturing calls for less inventory and a more real-time operations. How, then do manufacturers adopt and embrace this shifting scenario? The answer lies in becoming smarter and better through artificial intelligence.

Artificial intelligence can be applied to the entire lifecycle of manufacturing, right from problem identification to problem communication and then resolution. Automation is necessary to streamline repetitive tasks such as scheduling and rescheduling, planning and data tracking.

Prediction to preempt anomalies in operations goes a long way in bringing to notice aberrations and thereby avert critical catastrophes! Such AI-enabled systems must be instilled for material, machine, and equipment updates and also extended to the systems that they interact with, such as customer orders and supply-procurement.

Procurement and Supply Chain Management

The rise of AI in optimizing procurement function is astounding. Organizations today don't just measure what they spend, they are pushing the traditional approach of spend analysis to measure total value contribution to the business, taking advantage of both conventional and newly accessible data sources. How have they done this?

Artificial intelligence is adept at automating even complex tasks. From identifying new markets and tracking exchange rate volatility to managing risks and assessing the best suppliers, procurement as a corporate function is leveraging AI to streamline processes and improve decision-making. Spend analytics along with contract analytics have already gone through significant automation activities (processes of collecting, cleaning, classifying and analyzing an organization's expenditure data). As a result, the low hanging fruits of identifying areas where savings can be made and point to paths of greater efficiency have already been achieved, but this is only scratching the surface of the transformative powers of AI in procurement.

AI can bring in advanced strategic reasoning and strategic sourcing in to the mix. AI supplements such as cognitive procurement advisors (CPAs) and virtual personal assistants (VPAs) that use natural-language processing (NLP) and natural-language generation can further increase automation and efficiency in procurement. By facilitating creation of requisitions from free-form data like that contained in an email or PDFs (not just data entered into a requisition interface), AI can help bring more purchases under direct control. Speech recognition (such as Siri and Alexa) can change the requisition and approval process for busy and off-site workers. Also, using image recognition technology, employees may opt to send a picture of an item they wish to order or reorder, again simplifying the requisition process and making it likely that more spend is being actively managed. AI can play a significant role in reviewing and even generating and monitoring supplier contracts, automating a typically labor-intensive set of tasks. In addition, AI technologies can look for cost/price discrepancies, unusual order quantities or frequencies, compare contract data to orders and invoices, and a whole host of other patterns that will help companies detect potential fraud or errors, detect and predict purchasing patterns, and identify the top-performing trading partners.

Lastly, artificial intelligence is capable of "learning" your sourcing patterns and behaviors and making best supplier recommendations for a specific project or spend. It can even be used to predict market prices, identify and analyze new potential vendors, and help you evaluate the success of your relationships with existing suppliers.

If AI is to transform supply chain, it should be driven by issues that affect operations today. So, the first question to ask is "What needs to be improved in supply chains right now?" For example, typical retail/CPG supply chains carry 60-75 days of inventory, the average service level in the store is about 96%, with promoted item service levels much lower at the 80% range. The restaurant and casual dining industry, on the other hand, carries around 12-15 days of inventory with relatively high waste and high cost of goods sold.

The following are few challenges in SCM that AI can effectively address:

- *Hard to plan for demand*: Requires multiple iterations to arrive at a plan.

- *Excessive safety stocks and bullwhip effect*: Complex integrations and calculations at each step in the process and at each node in the supply chain network.

- *Supplier unreliability and transport network unpredictability*: Huge opportunities hidden in the network because they are locally sub-optimized.

In different shapes and forms, AI can help with all these problems. However for AI to offer optimal value in supply chain, it is important to ensure the following:

- *Access to real-time data*: The most important business imperative in SCM is to improve legacy batch oriented planning systems, but for that to happen we must eliminate the stale data problem. Most supply chains today attempt to execute plans using data that is days old, which results in poor decision-making that sub-optimizes the supply chain planning and requires frequent manual user intervention to stay current. Without real-time information, an AI supplement is just going to make bad decisions faster.

- *Access to varied data*: Unless the AI supplement can see all relevant constraints and bottlenecks due to external influence in the supply chain, the results will be no better than a traditional planning system. Hence, having access to all touch points related data, including external data, will help improve visibility into supply chain processes and inter-dependencies.

- *Last mile visibility and user engagement:* SCM goes through a number of process areas that are not integrated. The end consumer is the only consumer of your finished goods products. During the supply chain process, each function has their own goals and objectives, which one way or other stay aloof of the last mile impact. Hence, the primary goal of the AI engine must be to improve consumer service level at lowest possible cost but taking into account the entire supply chain touch points.

Data in a multi-party real-time network is always on a fast moving lane. In addition data variability and high latency are recurring problems. Hence, the AI engine can't afford to stay on a batch mode; it must be looking at the problems continuously and should learn as it goes on how to best set its own performance parameters to fine tune its abilities.

Significant value can only be achieved if the algorithm can make intelligent decisions and execute them. Furthermore, the algorithms need to execute not just within the enterprise but where appropriate, across the ecosystem of partners. This requires your AI system and the underlying execution systems to support multi-party execution workflows and, to a large extent, those workflows need to be automated.

Lastly, AI should not operate as a "black box." Users must get visibility to decision criteria, propagation impact, and they should also get a view of the issues that the AI system cannot solve. The users, regardless of type, should play an active role in monitoring and providing additional input to override AI decisions when necessary.

Shared Services Functions

Shared services functions in the corporate world are seen as cost centers, hence there is a natural tendency to partially outsource the support activities to drive cost optimization. Beyond the cost optimization, the service organizations need to shift focus to embed intelligence and automation in processes in order to offer higher-value services and improve service levels.

Many service organizations are starting to recognize the benefits of combining AI with robotic processing automation (RPA). They are using rules-based software bots to replace repetitive manual human activities adding flexibility, intelligence, and learning via AI. This approach combines the rapid payback of RPA and the more advanced potential of AI.

Conclusion

What are the keys to a successful AI strategy?

For ages, we have indulged in statistics, operations research, and automation to solve business problems and gain competitive advantage. For example, pricing is core to any business and the only goal is to increase the company's sales and margin. In simpler times, companies adopted different strategies for pricing that more or less stayed static for an extended period of time. To their pricing strategy, they had built-in scenarios, rules, heuristics, econometrics, and competitive intelligence. The more holistic and flexible the pricing strategy was, the more it served as a source of competitive advantage. However, today we are not in simpler times.

Ubiquity of digital in consumer lives, connected devices, the always online generation, and in general an increased level of awareness means consumers are no longer captive to one product or one service or one company. Today businesses are forced to focus on dynamic pricing (the right price to the right customer, at the right time, for the right product, through the right channel) to solve the same old problem statement—to increase sales and margin. This is where AI comes in.

Companies looking to achieve a competitive edge through AI need to reimagine their business models. They need to identify what machines can do better than humans and vice versa, develop complementary roles and responsibilities for each, and redesign processes accordingly. Executives need to identify where AI can create the most significant and sustainable advantage.

You need to ask four questions:

1. Is there a customer need?
2. How do we leverage technological advances?
3. Do we have access to data sources?
4. Can we decompose our processes?

First, understand the real needs of your customers. AI is an overhyped topic today, hence it always makes sense to return to the fundamental business questions. Do you have a view of your current or potential customers explicit or implicit unmet needs? Even the most disruptive recent business ideas, such as Uber and Airbnb, address people's fundamental requirements.

Second, incorporate technological advances. To do AI, you need to acquire and process new sources of data and push aggressive automation targets across your enterprise functions. The general availability of AI platforms and services can come to your aid, you can also take a different approach, of building your own AI infrastructure in house. The important thing is you need to put AI strategy on your board's agenda.

Third, create a holistic architecture that combines existing data with new or novel sources, even if they come from outside. The stack of AI services has become reasonably standardized and is increasingly accessible through intuitive tools. Even non-experts can use large data sets.

Finally, break down processes and offerings into relatively smaller and isolated elements that can be automated, instead of looking at a huge process flow and getting lost in the complexity. Once you have figured out a way to infuse AI supplements into these smaller processes, you can then do an end-to-end orchestration to optimize and automate the entire process flow.

For many organizations, these steps can be challenging. To apply the four questions systematically, companies need to have a clear picture of current and emerging capabilities of AI. A center for excellence can serve as a place to incubate technical and business acumen and disseminate AI expertise throughout the organization.

If you are the CDO of a company and are responsible for ushering in the AI transformative capabilities to your organization, what would you do? Besides the four questions that you must ask, you also need to focus on additional three areas, discussed next.

Develop a Clear Line of Sight to the Business Value

Start by assessing the relevance of AI from a business value in relation to specific operations and IT challenges.

Business value is an imperative. Many organizations become enamored with AI capabilities, but in the process they fail to determine the most strategic value drivers. You need to be certain about where to apply critical resources, such as data scientists and new solutions, that would benefit from AI and then firm up your plan to build capabilities where longer-term business outcomes are desired.

Expand your strategy with frameworks that will help you determine AI's applicability to business processes. Business process assessment frameworks establish a common language for describing your organization's existing business model. It also aids in assessing and proposing changes to individual components—improving cost structures, enabling data-driven revenue streams, or identifying new key partnerships where data and analytics can play a prime role. It also can you help identify changes to interrelated components that support potential extensive business model changes.

Harness Disruptive Potential in Customer Experiences

AI presents several opportunities for gaining insight, creating personalization, and enhancing the customer experience, which is one of the best opportunities for the use of AI and machine learning. Assessing its disruptive potential gives you the opportunity to engage customers in new ways, deepening your understanding of customer behavior and shaping the future of customer experience.

There are many opportunities to improve customer experience with AI, including developing customer insights and customizing their journeys, as well as predictive analytics for marketing. You'll need to leverage approaches such as journey mapping and outcome-driven innovation to identify unmet customer needs and opportunities.

Address Organizational, Governance, and Technological Impact

Prepare for the organizational, governance, and technological challenges imposed by AI. Lack of the necessary skills is often seen as a primary hurdle to AI adoption, so developing the necessary competencies will be critical. The obvious impact is with the development of data science skills and refactoring the CDO's organization to foster the creation and use of intelligence.

Many of the benefits of AI will come from the predictions rendered by machine learning. Yet, organizations are woefully ill-prepared to use these insights rather and evaluate and use probabilistic assessments of outcomes in decision making. This underscores the equal, if not greater, importance of developing a data-driven culture and the ability to "speak data" from a business perspective.

Using AI to gain insight into areas that humans can't underlies advancements in predictive analytics, natural-language processing, computer vision, image recognition, and many other displays of seeming intelligence. Numerous business scenarios certainly benefit from AI-generated insights and capabilities, but governing them may be a challenge due to a lack of transparency in how some of these approaches attain their results, a lack of processes to ensure quality results and appropriate use.

For example, it's possible that the same data with the same analytics may be governed differently based on use—one is ethically okay and the other is potentially not. The same may also be true for security, privacy, compliance, and retention.

To address these challenges, you need to develop a data-driven culture; be mindful of regulatory and ethical considerations; and steer clear of dangerous myths, all while fostering a learning laboratory for AI capabilities.

In this chapter, we discussed a variety of topics associated with AI, including some high-level pointers for developing your own AI strategy. In the next chapter, we discuss a very interesting question whereby the board asks the CEO to define the AI strategy.

References

1. https://www.bcg.com/publications/2017/competing-in-age-artificial-intelligence.aspx

2. https://sloanreview.mit.edu/article/ai-in-the-boardroom-the-next-realm-of-corporate-governance/

3. https://www.bcg.com/publications/2016/strategy-technology-digital-integrated-strategy-machine-using-ai-create-advantage.aspx

4. https://www.wired.com/story/amazon-artificial-intelligence-flywheel/

5. https://hbr.org/2017/10/how-ai-will-change-strategy-a-thought-experiment

6. https://psmag.com/news/artifical-intellgience-is-the-key-to-understanding

7. http://www.scmr.com/article/8_fundamentals_for_achieving_ai_success_in_the_supply_chain

8. https://www.gartner.com/smarterwithgartner/prepare-for-the-impact-of-ai-on-procurement/

Board to CEO: "What's Your AI Strategy?"

Dear CEO, artificial intelligence is your biggest threat and your biggest opportunity, so what's your AI strategy?

For many CEOs this question is a huge challenge. The topic of AI is so multi-faceted, hyped, and parabolic that it's hard to know where to start. What options do you have?

- You can acquire an innovative technology company that is focused on applying AI/ML to solve business problems and gain access to new products, offerings, and AI/ML talent.

- You can invest in a few early stage AI focused startups to be in the know about innovation happening in this fast-moving technology space.

© Soumendra Mohanty, Sachin Vyas 2018
S. Mohanty and S. Vyas, How to Compete in the Age of Artificial Intelligence,
https://doi.org/10.1007/978-1-4842-3808-0_3

- You can approve funding to build in-house AI capabilities and set up a centralized AI CoE to start exploring how to optimize your internal processes and how AI can be integrated into your products, offerings, and services. This option will be long drawn and you will need a strong vision, funding commitments, and a sense of purpose to attract and retain the best AI/ML talents.

- You can use outside consulting and services firms to fill the gaps in your existing business and IT teams to get the ball rolling.

These options are by no means mutually exclusive. On the contrary, you need to adopt an open mindset for your AI strategy—build/buy/partner/co-innovate could be the options—but you also need to measure/learn/experiment and then narrow down to what's working and what's not for your company.

While these options may give you some initial thoughts on creating and executing your AI strategy, there is more to it. How do you prepare for the unknown?

Now, back to the original question—what is your AI strategy? To get to the answer, first we will discuss why the board is asking this question. Then, we will discuss a conceptual framework to simplify AI so that you can give your board a thoughtful response. Finally, we discuss how you as CEO are going to engage your teams on the topic of AI and important considerations when formulating your AI strategy.

CEO: "Why Is My Board Asking This Question?"

So why is the board asking you this now? AI is neither an incremental tweaking of current IT systems and methods, nor is it about buying intelligence capabilities and platforms and making them part of your enterprise. AI brings in huge amount of transformative capabilities, cutting across the enterprise and impacting all roles across all levels. This can be very unnerving to most executives (even seasoned IT experts).

Look what happened just a decade ago! The DVD-rental company, the music company selling CDs, the brick-and-mortal travel planning and firms, the daily newspapers, the bookstore chains were all doing well and were set for long, profitable years ahead. The Internet happened and changed everything. Convenience shopping on the back of ecommerce became a way of life and well-established businesses lost relevance.

Today, businesses are staring at a similar inflection point. AI is taking us into uncharted territory. AI is helping machines converse, drive cars, perform physical work in hazardous conditions, perform delicate surgeries, predict what we would like to buy, diagnose health conditions, manage financial transactions, provide security, and much more. AI is everywhere. There are implications to our own jobs, our businesses, and quality of life, in general. To navigate this new territory effectively, business executives need to understand and consider a five-part AI game plan.

Apply the NFL Rule

The No Free Lunch (NFL) rule states that you can't have it both ways simultaneously—your AI cannot be good at both general intelligence and narrow intelligence at the same time. Don't get carried away by what you see in the press (investments by Google, Amazon, Apple, Microsoft, Tesla, etc. to develop general utility AI). You must focus on identifying and solving problems that are specific to your business and the markets you are in. Therein lies your strategic opportunity.

Embrace Openness and Transparency

It is important to understand and acknowledge that you can't do everything by yourself, hence collaboration and sharing is the key. Thousands of startups and others are trying to solve some very unique whitespace problems. Those solutions and ideas might not be in a disruption path for your markets at this point in time; however, they may become a reality really soon. Strategically, the best choice for you is to develop an inclusive ecosystem with startups, academia, and various AI labs and start leveraging the momentum to develop proprietary capabilities targeted to your specific market and your customers.

Evaluate Opportunities in the Connected World

The world is getting increasingly connected—ranging from fixed sensors to mobile phones to drones to smart wearables and highly conversational personal assistants. These edge systems are empowered to see, hear, understand, and react, thus creating enormous opportunities regardless of whether you are in a brick-and-mortar setup or highly digital setup. Naturally, the greatest AI value creation opportunities are there when you can effectively leverage these connected devices and come out with a unique business model.

Is this a market you should move into?

Focus on Augmented Intelligence

We are far away from creating human-like autonomous intelligence, so it's better to focus on areas where you can develop augmented intelligence offerings, both for your internal and external consumption.

As AI-driven automation starts affecting the knowledge-worker related skills, it will become essential to critically assess your approach to enterprise skills planning. Instead of taking conservative approaches to protect jobs, you need to evaluate and take few hard calls to make your human capital relevant (how to coexist in a human-machine integrated scenario) and to move up the value chain in the respective business processes.

Keep an Eye on Ethics and Cybersecurity

AI provides a consistent and objective approach to problem solving. Even so, it is not immune from biases. There have been several instances where the AI outputs have shown clearly biased results. Therefore, you must pay careful attention to the ethical, legal, and social concerns raised by AI.

CEO: "How Do I Respond?"

Let's start with some definitions of what AI is *not*. This is necessary because the media coverage of AI is not thoughtful and is often hyperbolic.

AI Is Not About Machines versus Humans

For the last few years there has been heightened speculations surrounding AI: humans will be replaced by machines. This is the completely wrong way of painting the picture of AI in the enterprise. The right vision is how we can leverage AI and augment human capabilities to do things that we had never imagined before. Even the recent media coverage of Google's DeepMind/AlphaGo victory over Lee Sedol was dramatized as machine achieving super intelligence and defeating humans at the games they had invented. A more realistic description could have been "machine learning from many humans' experience defeats a single human".

Machines have advantages that humans do not: speed, repeatability, consistency, scalability, and lower cost. Humans have advantages that machines do not: reasoning, experience, adaptability, and ability to handle a breadth of tasks. The smart thing to do is to find the right way to blend humans and machines, not replace humans with machines.

AI Is Not Always About the Best Algorithm

For many, the terms AI and algorithm hold the same meaning and importance. The best algorithm creates the best prediction. Facebook has the best newsfeed algorithm; Netflix has the best movie recommendation algorithm; and Google has the best ad placement algorithm.

Whether algorithms trump data or data trumps algorithms, the debate is on, but it is abundantly clear that algorithms are a necessary component of AI, but not the only component.

Many leading AI experts now have a view that lots of data and simpler data can do the trick as opposed to little or minimal data and complex algorithms. You get more out of gathering, cleaning, and understanding data than relying heavily on fancy algorithms. It's not quantity of data however, it's the quality that matters. A high quantity of data cuts both ways. Sometimes you have enough data that the answer is clear for even the least efficient methods, and with lots of noise. But other times, the sheer volume of data requires clever algorithms to make any sense at all.

Figure 3-1 summarizes the several building blocks for AI in the enterprise. The first two blocks are related to data and infrastructure; the middle block is all about algorithms; and the last two blocks are related to new skills, adoption of AI solutions, and the feedback loop.

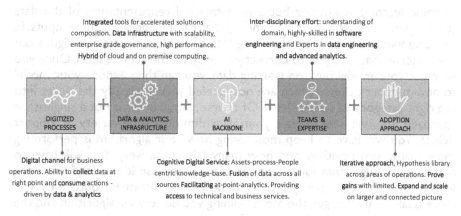

Figure 3-1. Building blocks for AI in the enterprise

So this brings us to a working definition of AI for the enterprise:

$$AI = \text{Training Data} + \text{Machine Learning} + \text{Feedback Loop}$$

This is the most important equation that the CEO needs to understand, if AI is to become a flywheel for the enterprise. Let's look at the individual components of this equation, taking a particular problem statement into account.

Let's suppose you want to do two things: 1) improve your service levels and 2) optimize your resources to eliminate activities that are repetitive. You would like to apply your AI solution to not only categorize support tickets by severity level automatically but also to create an intelligent knowledge base that will enable you to do event correlations, predict the ticket volumes, and automatically fix issues where you are 100% confident.

Training Data

Training data consists of any data (structured, unstructured, semi-structured, images, voice, and videos) that can help solve the problem you are going after. If the business process that you are trying to improve or automate with AI is already digital, then the data that is required to complete that process may already be available. If part of the data is labeled, meaning the data has certain indicators that you can use to correctly classify the outcomes, then you are in a good shape to start training your machine learning algorithms. In the context of our example, if we have data that besides having information about the support ticket also has an indicator showing the "severity level," then we call that data set labeled.

Machine Learning

Machine learning is nothing but a mathematical representation of the data to help you arrive at a predictive model that can be applied to new inputs. In our example this means, by observing lots of support tickets, you figure out the patterns that lead you to correctly determine the service level. Once you have got your model built on training data, you can then expose your model to new support tickets (which are not labeled yet) to classify the ticket into the right category of service level. Of course, this process is not a single-shot approach. You will have to train your algorithm on a wide variety of support tickets. You will have to keep monitoring how your algorithm is performing (how many false positives and how many false negatives). You will have to penalize your algorithm for wrong predictions because it may lead to wrong fixes getting applied in live systems with potentially disastrous outcomes. The more quality data you get, the more refining you do to your algorithm and the better prediction capabilities you achieve.

To achieve a high degree of prediction, you have to put in a sustained effort to acquire diverse set of data including process maps, system behaviors, user journeys in the system, possible root causes, a large corpus of scripts that can fix any kinds of system issues, troubleshooting tips and tricks, and any other information that can come benefit your algorithm. This diverse data set, if curated properly and maintained well, will become the intelligent knowledge base for your enterprise.

One of the advantages of algorithms compared to humans is the true reflection of their own confidence level. Humans are notoriously overconfident about their own reasoning capabilities and judgments, which is people dependent. Two support engineers looking at the same support ticket will rarely arrive at the same conclusion if left to their judgment, unless they are bound to follow a standard operating procedure that has steps outlined to an excruciating level of detail. In contrast, algorithms will be always consistent in what they predict (right or wrong).

Feedback Loop

When algorithms start training on data, they start from no experience. They have no knowledge about patterns hidden deep inside the data. They can make predictions but someone needs to validate those predictions. This is where human experience comes into the picture, by providing feedback, reviews, and sometimes even overriding the machine predictions. This is the critical third component in the equation.

So far we have been highlighting how the prediction technology will give impetus to more automated decision making, but at the same time we are also saying that the feedback loop is important, that human experience and involvement during the learning process are important. Aren't these views contradictory? There could be scenarios where the algorithm will go blank. The reasons could be many—the new data it sees is completely different from the data it was exposed to during the learning process, such as certain rarely occurring or in-frequent scenarios it encounters for the first time which was not there in the training data. The machine then objectively assesses the data and comes out with predictions that are highly accurate from the machine's point of view but it are not aware of the numerous business rules baked into the business processes and applications. These are the primary reasons why human involvement during the learning process is extremely important. Unless the machine is validated on its predictions, it will never reach the level of confidence where you can pretty much automate the entire decision-making process.

CEO: "How Do I Engage My Leadership Team?"

Now that you have an understanding of how to apply the AI framework to your business, it's time to start engaging with your senior leadership team. It is time to strategize and push a blended model that takes advantage of the speed and scale of machines to address the well-defined but repeatable tasks, while the humans handle the higher order judgment tasks.

Focus on Outcomes Not Technologies

First, speak the language of outcomes that matter to your senior executives, not the techno speak that comes along with any new disruptive technology.

A good way to engage your leadership team is by asking a few questions:

- Where can we infuse responsiveness and prediction to better deliver customer experience, launch new products faster, or improve our own operational efficiencies?

- Which activities that we are doing today can be done faster, better, cheaper, and in an automated way?

- What are the new things we wanted to do but could not do due to various constraints (cost, resources, or technology)?

- What are the things that we learn too late and therefore cannot act in time?

- What are the things that we ought to know could happen—good or bad?

For example, your SVP responsible for customer service is concerned about the decreasing trend in "first time right resolution". Your president of sales is concerned about the dip in "cross-sell/up-sell" effectiveness and is linking the issue to bad customer service experience. Your VP in charge of product strategy is keen to understand the market pulse, but is unable to do so because you do not have the means to collect and analyze the relevant social media data, search queries data, and brand equity data. Without that data, you can't really understand what customers are looking for and talking about.

Explaining AI

Once you've identified the outcomes that matter to your senior executives, you can establish how AI can help the business meet its business objectives. Here is where our working definition of AI will come in handy.

For example, your VP products' strategy and your CMO are keen to understand the customer pulse so that they can design better products and launch effective campaigns and gain more market share. For this to happen, you need AI capabilities to collect and analyze social media data and do sentiment analysis at scale.

Now, stepping through the working definition of AI = Training Data + Machine Learning + Feedback Loop:

Training Data. You need various types of data: tweets from Twitter based on the #hashtags and @mentions that matter, as well as Facebook conversations around your company, brand, and product experiences. Most importantly you need to know what people are saying about competing products, other publicly available data related to competitors strategy, and any innovations that promise to deliver better customer experiences or make your product obsolete. Most of this data is streaming and real-time so you will need to figure out how and where you will store all this data and for how long. This data also won't be of high quality and isn't labeled to begin with, so you will have to define a process through which you can clean the data, filter out noises, define a sentiment scale you want to use, and deploy a few subject matter experts to help with labeling.

Machine Learning. Now comes the time to talk about algorithms. With the curated data set you have, you will have to define what type of prediction you want to do. Several types of predictions are relevant to the problem you are trying to solve:

- *Classification*: Looking at the sentiment data, you have collected you need to develop cases like "positive sentiment" or "negative sentiment". Classification algorithms will help you scan through vast amounts of data to arrive at classes of outcomes.

- *Segmentation*: If your intent is to identify clusters of people who have expressed positive sentiments and are most likely to respond to your offers, then clustering algorithms will come into play.

- *Regression*: If your intent is to arrive at a potential number (could be a revenue upswing or the number of products you will be able to sell by market/geography), then you will need to apply regression algorithms.

- *Forecasting techniques:* If you want to understanding various temporal and seasonal or other external influences hidden within the data and want to create projections (could be about the effect of campaigns to sales), you need to turn to forecasting algorithms.

There are many more variants of algorithms at play, including deep learning and neural nets. However, for the sake of brevity and to keep things simple, we have taken the examples of simple algorithms here.

Feedback Loop. No matter how brilliant your algorithm is, you have to make the prediction outcomes actionable. The only way people will start using the prediction recommendations is if they understand the why and how of the output. In short, your algorithm can't behave as a black box. The output needs to be explainable and people need to have flexibility in validating the outputs and as a consequence help in refining and enriching the prediction capabilities.

Once your senior executives understand that it is not going to be the machine-way all the way, they will be much more appreciative of how AI can augment human skills in the enterprise.

Evaluating AI Platforms

It is not surprising to see enterprise IT landscape littered with many vendor products powering business applications and IT management applications. A natural reaction from enterprise IT architects and enterprise standard groups is to reduce entropy in the IT landscape. This means introducing tools and technology and another long process of evaluations, commercial negotiations, contracts, and certifications. Managing vendor risk and safeguarding the enterprise from vendor lockdown is on the top of every IT leader's agenda. AI, being a new and fast evolving technology, also has its fair share of technology architecture and components (whether on-premise or on-cloud). The challenging part is that even your seasoned IT leaders will be struggling to articulate where this technology is moving.

Your IT leaders will need a helping hand to focus on what is relevant (there is already a lot of tall claims from vendors about their AI platforms capabilities) than going through a never-ending phase of evaluating vendors. You should surely seek support from your CIO/CTO colleagues to do an extensive evaluation of tools available in the market. You should also ask them to profile the startups to learn what they are trying to achieve, many times a collaboration approach with the startups can do the trick. This is commercially less taxing for you and motivating for your teams, as they will get involved in cutting edge technology and innovative ideas.

The following three factors are worthy of an extra look, specifically for vendors claiming to deliver AI solutions.

Single Domain versus General Purpose

Some vendors have gone deep into one specific domain (pharma clinical trials or hyper-personalization for retail multi-channels or fraud detection in the financial crimes scenario). Some vendors have chosen to specialize on technology capabilities (IoT, image recognition, NLP, and deep learning), while many others have chosen to build a general-purpose platform (you can ingest any kind of data, you can generate insights by using popular machine learning techniques, and you can distribute you insights through multiple consumption channels). These vendors don't necessarily specialize in any specific AI subfield like text, images, audio, video, deep learning, etc.

It is difficult to say which approach is better than the other. It all depends on the types of problems you are going after. However, instead of choosing horses for courses (multiple tools for multiple use cases as they deem to be fit), the natural choice for the enterprise is to rationalize the number of vendors they manage so there is a slight bias toward a general purpose platform, if (and it's a big if) that general purpose platform can meet the business needs of different functions.

Black Box versus White Box

Algorithms have attitude too. Some are simple and do their job quietly and are easy to interpret (decision trees, linear regression, logistics regression, Naïve Bayes, etc.). Others (especially the new generations of algorithms based on neural net) are complex, versatile, and extremely difficult to comprehend. Many of these algorithms cannot explain their results even to the data scientists who built them, let alone to end users. They operate like black boxes, in which you can't really inspect how the algorithm is accomplishing what it is accomplishing. Several recent examples—like Google Photo's mislabeling of an image of a black couple as gorillas, and Amazon bookseller bots that bid against each other until a book's price exceeded $23 million—create a trust issue in our minds.

What do we really mean by explaining the machine learning algorithms? With respect to the predictions we get from algorithms, we typically assume that the algorithm has done its homework, using all the right reasoning, and we are basically asking the algorithm to be transparent enough so that we can see the reasoning process it used to arrive at the prediction.

Choosing between a black box and a white box is an important decision for your organization. There are obvious reasons why anyone would opt for a white box approach, but there are implications too. Too much fiddling around with the inner workings of the algorithms to make it transparent can make the algorithm run slow and can create bottlenecks in your automated decision-making journey because of too much human involvement even for simple prediction-related problems.

Point Solution versus Integrated Platform

Commercially viable AI platform needs the three critical components—training data, machine learning, and a feedback loop—to be integrated. As long as you are getting all three capabilities, you should be good. Some vendors focus on features such as the ability to connect to any type of data sources and systems, including devices, and then offer a completely open canvas to do your ML. This is done by programming intensive languages like R and Python or by leveraging drag-and-drop features like Azure ML, Spark ML, AWS ML, etc. Other vendors focus largely on data engineering aspects through Spark, Scala, and Python and then leave the ML to be done by your choice of technology.

It is your choice, how you want to go about it.

CEO's New Management Principle: Management By AI

All said and done, every CEO wants to leave a legacy behind. What does that mean? It is all about leading and inspiring people and building a future ready enterprise, so it is quite natural that they don't like delegating critical business decisions to smart algorithms and don't like clever code recommending management tips. In some of the world's most successful companies of our times (Google, Amazon, Alibaba, and Facebook), algorithms are increasingly playing the decision-making role—from the boardroom to sales and marketing to operations to hiring talent. CEOs should add one more management doctrine into their best practices—empowering algorithms. Elite MBAs (management by AI) are slowly becoming the new normal.

This is a tricky situation, because going too far and leaving everything to the algorithms to manage will create unrest within the rank and files of the company, whereas doing too little with algorithms is not a choice anymore. Therefore, the right balance of talented humans and elite MBAIs is required. The other issue of importance is to define clear lines of authority and accountability—the whats, hows, and whys of coexistence and delegation. CEOs need to establish a framework in the lines of the RACI matrix (Responsibility, Accountability, Contribution, Information) to clearly articulate who is in charge of doing what and when talented humans must defer to algorithmic judgment and when algorithms must require human validation.

There is another aspect of "feeling insecure" that makes the whole scenario of human-machine collaboration harder to implement. Leaders who had shown passion and commitment to, let's say, automating a complex business process to improve the supply-chain function or a geography specific operations that was highly inefficient, would flinch at the prospect of deep-learning algorithms dictating their business strategies, new market exploration, and capex planning. The implications of the algorithm's success scare them more than the risk of failure.

For example, your data science teams, in collaboration with your domain experts from procurement and supply chain divisions, developed an algorithm that, by all measures and simulations, has demonstrated that you can save hundreds of millions and will enable your company to respond 10 times faster to market dynamics than your existing batch-oriented processes. Why wouldn't you trust this brilliant AI solution? This is where the challenge lies.

When computers became ubiquitous and information technology became the backbone of business, CEOs appointed CIOs and CTOs to help companies organize their technology landscapes, information management architectures, and business applications. As technology matured and especially when every business became digital, CEOs appointed CDOs (Chief Digital Officer) to implement and govern their company's digital strategy. Time has now come to say the same thing about the transformative role that AI can play in the company's future readiness. CEOs should seriously think about appointing a CAIO (Chief AI Officer).

AI is still maturing, so it is reasonable to accept that not everyone in the C-suite would understand the benefits and implications of AI in an enterprise landscape completely. But, think about it, if your company is into the business of products/services and deals with large amounts of data (whether internally or externally, it does not matter), then there is a good chance that AI can be used to transform that data into value. Companies that have data but lack deep AI knowledge, just like you had previously appointed CIOs/CTOs, now need to appoint a Chief AI Officer. If you already have a Chief Digital Officer or Chief Data Officer, or if you have some forward-thinking CIO, you may expand their role to also look into AI.

More mature organizations have a Chief Data Officer with a centralized data science function, but the more common model seems to be a decentralized model with data scientists sprinkled across an organization inside different functions.

In the decentralized scenario, there will inevitably be a broad spectrum in the skillsets of the data scientists in a single company. Some will have advanced machine learning backgrounds, while others will have progressed to the data science role from being a software engineer or a data analyst. Those with the advanced machine learning skills probably won't be satisfied with a black box approach, where it's not transparent how an algorithm works and they can't fine-tune the parameter weightings. Bear this in mind when considering solutions that are black box only.

AI itself is not a product or a business unit by itself; it is a foundational technology that can help existing lines of business and create new products or lines of business. Thus, to successfully usher AI into an enterprise requires a great deal of understanding of your current business and demands working with diverse business units or functional teams. You need a leader who can exhibit entrepreneurial traits. With AI evolving rapidly, your teams will need to keep up with changes. It is less important that they be on the bleeding edge of AI. What is important is to choose an evolutionary path. Your teams of data scientists, data engineers, and executives need to work cross-functionally and have the business skills to figure out how to develop and implement AI solutions that are relevant to your enterprise and business goals.

The chief AI officer needs to have a good technical understanding of AI and your company's data infrastructure, must have the ability to work collaboratively across business functions and executive leaders, and most importantly should have credibility to attract and retain AI talent.

CEO Mandate: Democratize AI

The other most important aspect in the CEO's AI strategy is to make AI technologies widely available throughout the organization, especially to those who have loads of experience managing and running your business functions but are at sea when it comes to using the latest technology led decision making. Call it the "democratization of AI". To accomplish this goal, the company needs to do several things: develop their own AI platform, conduct extensive design thinking sessions across business functions, and re-skill the workforce to understand how prediction technologies work and how they can coexist in an increasingly human-machine world.

In the longer run, your workforce will get enough exposure and know-how to start developing their own AI applications to do their job faster, better, and cheaper. For example, your marketing team will be able to connect a widget of natural language processing together with other components to create an app for gathering and analyzing unstructured data from social media data. Or your sales team will be able to analyze a huge customer data set to predict which prospects are most likely to convert to customers or uncover insights about customer behavior and improve the design of different products and offerings, thus increasing the ability to cross-sell/up-sell.

Democratization of AI will also allow your ecosystem partners to co-platform, bringing in strong collaboration capabilities and simplified process orchestration that cuts across organizational boundaries. Think of how spreadsheets helped democratize data analysis, enabling even mom-and-pop shops to perform invaluable "what-if" analyses.

It is clear that the "democratization of AI" is a strategic undertaking that all enterprises need to do to create a sustainable competitive advantage.

Conclusion

AI is a longer-term trend that will impact all aspects of our lives at home and at work. We are moving to a world that's a big semantic web where everything is connected with everything. Purely talking in algorithmic norms, our location is an attribute, our sentiment is an attribute, our facial expressions is an attribute, and our voice is an attribute. The super convergence of cloud, mobile, social media, Big Data, AI, and now blockchain are enabling companies to aggressively explore, innovate, and launch disruptive businesses models and super smart offerings in the consumer-to-consumer and business-to-business sectors. In this context, the board asking the CEO about an AI strategy is apt. Enterprise AI is about building, deploying, and managing systems of intelligent engagement. It is not about mere automation of manual work. Thus as a CEO, your task is cut out for you. You need to have a clear vision:

1. *Establish AI as a business capability*: AI is too important to be left to few experimentations or one-off AI proof of concepts or the IT teams to develop know-how. The point is not about publishing an annual report where you can also say that you are exploring AI, like many others do. CEOs should be asking: what is the art of the possible with AI?

2. *Democratize AI within the company*: Incrementalism will not transform companies. Adopt a "dot versus bubble" analogy to arrive at how your company should be approaching AI use cases not only to differentiate in the market but also to democratize AI within the company.

3. *Adopt fail-fast learn-fast methodology*: Adopt a 10/10/10 methodology: 10 hours to identify use case, 10 days to build the data pipeline, and 10 weeks to go live with AI capability. CEOs must encourage a culture of learning and experimenting. This requires an agile methodology to go from concept to capability in fewer than 90 days.

4. *Drive and manage AI projects as a top-down initiative*: The leadership has to treat AI as a business imperative that is driven from the top-down, not a technology know-how drive and experimentation type of projects driven from the bottom-up.

5. *Build an ecosystem to leverage*: The hyper-convergence of the big six technologies means that companies need an ecosystem of partners to tackle all of the key technologies.

What if your board doesn't ask you this question? If your board of directors is not concerned about the company's AI strategy to drive new business models or new products and services, maybe you need to consider changing the composition of the board! In this fast-paced changing business and market scenarios, there are no second chances. You absolutely need your board to understand and play a critical role in questioning your strategy, technology investments, and people reskilling initiatives that your company's future depends on.

In the next chapter, we discuss fundamentals of AI (slightly technical aspects of AI) and the associated apprehensions, despite its obvious advantages.

References

1. https://hbr.org/2017/03/the-trade-off-every-ai-company-will-face

2. https://www.mckinsey.com/business-functions/digital-mckinsey/our-insights/digital-blog/what-every-ceo-needs-to-know-to-succeed-with-ai

3. https://hbr.org/2017/08/a-survey-of-3000-executives-reveals-how-businesses-succeed-with-ai

4. https://www.cio.com/article/3167964/artificial-intelligence/building-the-predictive-enterprise-what-cios-should-tell-ceos-about-ai.html

5. https://www.crowdflower.com/ceo-cio-whats-ai-strategy/

6. https://hbr.org/2017/01/4-models-for-using-ai-to-make-decisions

7. https://hbr.org/2016/11/hiring-your-first-chief-ai-officer

Inside the Black Box: Understanding AI Decision Making

AI's ability to keep improving its predictive capabilities just by learning from the data and without significant involvement from humans to explain exactly how to accomplish the tasks is a big deal. Why?

Two reasons. First, we have a lot going in our brains but we can't explain exactly how we're able to reason, differentiate, adapt to changing environments, and use experience to solve problems we have never encountered before. Until AI, our attempt to come closer to human-level thinking involved complex "If-then-else" style of programs to solve problems. This approach is neither scalable nor adaptive to changing conditions.

© Soumendra Mohanty, Sachin Vyas 2018
S. Mohanty and S. Vyas, *How to Compete in the Age of Artificial Intelligence*,
https://doi.org/10.1007/978-1-4842-3808-0_4

Second, AI systems are excellent learners. They learn from data, they learn from validations done by humans, and they learn when they are penalized for wrong predictions. There is very little coding involved to give AI these kind of learning capabilities.

The term artificial intelligence (AI) implies a machine that can reason. A more complete list of AI characteristics is:

- *Reasoning*: The ability to solve problems through logical deduction

- *Knowledge*: The ability to represent knowledge about the world

- *Planning*: The ability to set and achieve goals

- *Communication*: The ability to understand written and spoken language.

- *Perception*: The ability to engage with the world through vision, sound, and other sensory inputs

There are three major classes of AI based on their problem-solving capabilities.

Artificial Narrow Intelligence (ANI)

Also known as "weak AI" because the problem definition is narrow. The AI that beat the world chess champion is an example of ANI. It is very good at playing chess, but that's the only thing it does. If you asked the AI specialist in chess to give you driving directions from point A to point B, it would look at you blankly. There could be another AI specialist to help you with driving directions, but it can't help you with chess.

You can imagine that ANI has a specific purpose and has intelligence that is confined to one area, and mostly one or a few specific goals. Intelligence here is more often defined by the ability to work alone rather than (current/ gradual) self-improvement.

At the moment, we use hundreds narrow AIs in real life. Siri, Google, Facebook, self-parking and self-driving cars are more sophisticated (intelligent) narrow AIs in action. You can also find weaker AIs, like watches that can change time zones based on your location or automatic temperature control in your house based on weather conditions outside.

Here are a few more examples of ANIs in our daily lives:

- Modern cars are full of ANI systems, from the automatic anti-skid mechanism to alerting you about sudden loss of tire pressure to giving you an indication that there is another vehicle in your blind spot. Self-driving cars, on the other hand, is an example of sophisticated ANI system that not only engages with all the mechanics of a car and drives the car for you but also interacts with the outside environment to make decisions on your behalf—when to stop, when to maintain a safe distance, when to change lanes, etc.

- Your smart phone consists of a lot of weak ANI solutions, from a shopping app that recommends purchases to real-time driving directions from point A to point B, to getting weather updates, to listening to your preferred music list, to engaging in a conversation with your phone, and many more apps that organize your everyday activities.

- Microsoft outlook or Google's Gmail are intelligent ANIs. They come pre-built specialized in classifying what's spam and what's not. They are continuously working in the background observing your actions and preferences to classify what you consider as important, junk, and spam.

- Amazon's "People who bought this also bought..." recommendation system is a continuously learning ANI system that gathers information from the buying patterns of millions of customers and then personalizes that information to influence you to buy more things.

- Google offerings like search, Google translate, and Google news are examples of sophisticated ANI systems specializing on content. The same goes for Facebook's newsfeed.

- There are a few more examples of ANI systems even though they specialize in solving only one problem but they go really deep into that domain. Security systems, automated manufacturing, algorithmic trading in capital markets, expert systems to help doctors make diagnoses, and, most famously, IBM's Watson, which contained enough curated information to beat the most prolific *Jeopardy* champions.

Artificial General Intelligence (AGI)

Also known as "strong AI," this is an AI system that can perform any intellectual task that a human being can. We are far from creating AGI systems because intelligence is a complex thing and it largely involves reasoning, ability to adapt, and mental capability. Thus, if AGI has to exist, it has to do everything that humans do—use strategy and make judgments under uncertainty, concentrate knowledge, including common sense, plan, learn, communicate in natural language, and then combine all these higher-order skills to solve problems and attain goals. This is a tough ask!

Artificial Superintelligence (ASI)

Everything that goes beyond AGI is ASI. The ASI systems are smarter than the best human brains in practically every field, including scientific discovery, creativity, general knowledge, and social interaction.

As of now, we have only reached the lowest class of AI. The road from ANI to AGI to ASI will require tremendous amounts of technology evolution. For sure, it will be an exciting journey, but a journey that will challenge the very existence of humans.

Categories of Learners

Machine learning algorithms "learn" to make predictions. The approach they take to learn is solely driven by the type of problem you want to solve.

Supervised Learning

You have a target or a class to predict. For instance, let's say you want to predict "who is going to attrite?". Then your model will be trained on historical data, where you have labeled attrition status against each customer. The algorithm looks at various patterns hidden in the data to map to the customer status (attrition = yes or no). This process of mathematical abstraction of learning from data to arrive at the desired outcome is called *supervised learning*, because it knows what to learn.

The key principles to know are:

- The training data set contains the predictors and the output you want to predict.

- The learner will use the training data set to determine a map between the predictors and the output.

- Once the learner learns the map, it can then be used on new data with some accuracy.

Here are a few examples:

- Input (Voice Recording) – Output (Transcript) – Application (Speech Recognition)

- Input (Photograph) – Output (Caption) – Application (Image Tagging)

- Input (Store Transaction Details) – Output (Is the Transaction Fraud?) – Application (Fraud Detection)

- Input (Recipe Ingredients) – Output (Customer Reviews) – Application (Food Recommendations)

- Input (Past Purchase Data) – Output (Future Purchase Behavior) – Application (Recommendation System)

- Input (Vehicle Location and Speed) – Output (Traffic Flow) – Application (Traffic Management)

Unsupervised Learning

You are in general looking for patterns and have no outputs to map to. For example, you want to group your customers according to the type of products they order, how often they purchase your product, their last visit, etc. This process of arriving at a cluster with homogenous behaviors grouped together is called *unsupervised learning*, because it does not know what to learn. However, it observes the patterns in the data and automatically finds ways to find distinct groups in the data.

The key principle to understand is:

- Unsupervised learning does not try to map into a predefined output, rather it explains how the data is organized.

Here are a few examples:

- A clustering algorithm, such as K-means or hierarchical clustering. These algorithms try to create clusters (similar characteristics belong to one cluster).

- Dimensionality reduction such as PCA. These algorithms try to find the best representation of the data with fewer dimensions.

- Anomaly detections. These algorithms try to find outliers in the data.

Reinforcement Learning

You want to attain an objective but you are not given any specific set of instructions or prior lessons about how to achieve the objective. In reinforcement learning, the learner is not told how to do the tasks. However, the learner is explicitly informed about which actions will lead to maximum reward.

The key principles to know are:

- Reinforcement learning tries to acquire new knowledge and maximize reward at the same time. This approach is also known as the "exploration vs. exploitation trade-off".

- The learner follows the Markov Decision Process framework, where it has to take an action (A) to transition from a start state to an end state (S).

- For every correct action it takes, it gets a reward (R). Otherwise it gets penalized. This sequence of actions defines the policy (P) and the corresponding rewards (or penalties) define the value (V). The learners' task here is to maximize the rewards by choosing the correct policy.

Reinforcement learning is not exactly supervised, because it does not rely strictly on what output it has to map to. It's not unsupervised learning either, since it knows upfront what is the expected reward for the right actions.

Categories of AI Problems

AI problems can be classified into 10 broad categories.

- *Domain expert:* The AI system must acquire a body of knowledge in a specific field and then gain enough reasoning and inference capabilities that it can act as an expert in that field. The key is here to get access to a huge corpus of data concerning a particular field.

- *Domain extension:* The AI system leverages a body of knowledge and extends the capabilities to correlate different things within that body of knowledge to suggest new insights or patterns that were not discovered earlier.

- *Complex planner:* The AI system serves the role of an intelligent orchestrator that not only integrates a variety of complex data but also generates insights. One example is the use of AI techniques to optimize logistics where, due to multiple hops and parties involved, including fleets, warehousing, routing, and distributions, you are looking at large and complex data sets in which human beings cannot detect patterns but a machine can do so easily.

- *Better communicator*: The AI system provides a conversational mechanism, including natural language processing, natural language generation, interpreting emotions, tailoring responses, language translation, etc.

- *New perception*: The AI system interacts with environments to determine the next best actions enabling new services such as autonomous vehicles.

- *Enterprise AI*: The AI system is primarily tasked to automate activities that are repetitive in nature where there is enough knowledge base available to define input-output mapping accurately.

- *Making ERP intelligent*: Businesses run on ERP systems. However, they lack in intelligence. AI systems are embedded in the various workflows in the ERP system and are tasked to make smart assists to users who are stuck in the workflows or need assistance to carry out their tasks efficiently. The AI systems are trained on past issues, conflict resolutions, best practices, possible remediation measures, etc.

- *Making data stores intelligent*: Enterprises have many data stores. Some are managed centrally (data warehouses) to serve analytical needs, some are associated with specific business applications and stay siloed, and many more are created as a result of point-in-time needs of users. In addition, there is quite a large amount of information scattered around in unstructured formats (logs, emails, documents, images, videos, call transcripts, etc.). AI systems can play a role in cutting across all these different and distributed data stores to not only collect metadata (data catalogs, business glossary, and technical glossaries) but also to serve as a means to provide meaningful insights, letting the data stay wherever they are.

- *Super long sequence pattern recognition*: The current scenario of AI systems doing a fabulous job in predicting the outcomes that are perhaps applied to one single problem. There are many large, complex real-world problems that require a sequence of predictions to feed into each other to solve the problem. Such AI systems will not only have their own memory but also will have advanced synchronization capabilities to self-tune and adapt.

- • *Advanced behavior analysis*: AI systems are already doing sentiment analysis but they apply a limited interpretation capability, primarily using vocabularies and grammar constructs. These AI systems can evolve to use multiple areas of understanding human behavior, such as observing body language, facial expressions, choice of words while communicating, tone of interaction, etc. to enhance sentiment analysis.

ML Primer for Managers

The best way to understand the engineering aspects of AI and algorithms is to refer to the definition provided by Tom Mitchell.[1]

> *"A computer program is said to learn from experience E with respect to some task T and some performance measure P, if its performance on T, as measured by P, improves with experience E."*

So if you want to predict, for example, credit worthiness of a new loan applicant (task T), you run a machine learning algorithm on past data about similar profiles (experience E) and, if it has successfully "learned," it will then correctly predict the credit worthiness of the new loan applicant (performance measure P).

The answer to the question, "What machine learning algorithm should one use?" is always "It depends." Choosing the right algorithm is a combination of the type of problem you are trying to solve, business goals, experimentation techniques you want to apply, and how much time you have on hand to validate your algorithm. It depends on several factors: data size, data quality, and data diversity. Additional considerations include your tolerance for error and what you want to do with the answer. Even the most experienced data scientists cannot tell you which algorithm will perform the best before trying them.

We have picked top 10 widely used machine learning algorithms to get you initiated into the algorithm-jargons. It also means that there are lots of algorithms not listed here.

Naïve Bayes Classifier

The Naïve Bayes classifier is based on Bayes' theorem and allows us to predict a class/category, based on a given set of features, using probability.

---

[1] Tom Mitchell, *Machine Learning* (McGraw Hill, 1997).

Naïve Bayes can be applied only if the features are independent of each other. If we try to predict a flower type by its petal length and width, we can use the Naïve Bayes approach, since both those features are independent. For instance, spam filtering is a classifier that assigns a label "Spam" or "Not Spam" to all the emails.

K-Means Clustering

K-means clustering is a type of unsupervised learning that's used to categorize unlabeled data. K-means is a non-deterministic and iterative method. The algorithm follows a procedure to form clusters that contain homogeneous data points.

The value of k is an input for the algorithm. Based on that, the algorithm selects k number of centroids. Then the neighboring data points to a centroid combine with its centroid and create a cluster. Later a new centroid is created within each cluster. Then data points near to new centroid will combine again to expand the cluster. This process is continued until the centroids do not change.

KNN (K-Nearest Neighbors)

KNN algorithm can be applied to both classification and regression problems. The algorithm stores all available cases and classifies any new cases by taking a majority vote of its k neighbors. The case is then assigned to the class with which it has the most in common. A distance function performs this measurement.

Support Vector Machine

This is a type of supervised learning that essentially filters data into categories by finding a line (hyperplane) that separates the training data set into classes.

As there could be many ways to create linear hyperplanes, the SVM algorithm tries to maximize the distance between the various classes using a margin maximization technique. SVM is a method of classification in which you plot raw data as points in an n-dimensional space (where n is the number of features you have). The value of each feature is then tied to a particular coordinate, making it easy to classify the data. Lines called *classifiers* can be used to split the data and plot them on a graph.

Compared to other classification algorithms, SVM gives better accuracy on the training data. SVM algorithm does not make any strong assumptions on the data and does not overfit the data as well.

A Priori

This is a type of unsupervised machine learning that generates rules-based inference from a given data set. Association rule implies that if an item A occurs, then item B also occurs with a certain probability. Most of the association rules generated are in the IF_THEN format. There are certain derivation principles this algorithm takes into account:

- If an item set occurs frequently then all the subsets of the item set also occur frequently.

- If an item set occurs infrequently, then all the supersets of the item set have infrequent occurrence.

Many e-commerce giants such as Amazon use A Priori to draw inferences, like which products are likely to be purchased together and which are most responsive to promotion. Similarly, by observing the frequency of words used together in a search query, Google magically suggests recommendations for the next set of words, even when you have not stopped typing.

Linear Regression

The Linear Regression algorithm helps us understand the relationships between two continuous variables and how the change in one variable impacts the other. One of the most interpretable machine learning algorithms, it's easy to explain to others and requires minimal tuning.

The relationship is established between independent and dependent variables by fitting them to a line. This line is known as regression line and represented by a linear equation $Y = a * X + b$.

In this equation:

- Y is the dependent variable
- a is the slope
- X is the independent variable
- b is the intercept

The coefficients a and b are derived by minimizing the sum of the squared difference of distance between data points and the regression line.

Logistic Regression

Logistic regression is used to estimate discrete values (usually binary values like 0/1) from a set of independent variables. It helps predict the probability of an event by fitting data to a logit function. Here the outcome of the prediction is not a continuous number but largely a classification related output.

Based on the nature of a categorical response, logistic regression solves three types of problems:

- *Binary Logistic Regression:* Applicable when the categorical response has two possible outcomes, i.e., yes or no.

- *Multi-Nominal Logistic Regression:* Applicable when the categorical response has three or more possible outcomes with no ordering.

- *Ordinal Logistic Regression:* Applicable when the categorical response has three or more possible outcomes with natural ordering.

Decision Trees

A decision tree is a graphical representation that uses a branching methodology to plot a path leading to decisions. In a decision tree, starting from the root, each internal node undergoes a test on the response variable—each branch of the tree represents the outcome of the test and the leaf node represents a particular class label, i.e. the decision made after computing all of the attributes. The classification rules are represented through the path from the root to the leaf node.

When the response variable is categorical in nature, we apply classification tree algorithm. When the response variable is continuous or numerical in nature, we apply regression tree algorithm.

Random Forest

Random forest as a collection of decision trees. It uses a bagging approach to create a number of decision trees, taking random subsets from the original data set. In this ensemble learning method, the output of all the decision trees in the random forest is combined to make the final prediction. Each tree "votes" for that class. The forest chooses the classification having the most votes (over all the trees in the forest).

Dimensional Reduction Algorithms

Not all data (attribute) is equal in terms of its value and importance in arriving at a prediction. However, every single raw data point, if curated well, can deliver tremendous value. Thanks to Big Data technologies, we now have the ability to store, manage, and deal with data sets consisting of thousands of variables. The challenge is in identifying the variables that will have the most impact on your prediction.

Dimensionality reduction algorithms like PCA, Decision Tree, Factor Analysis, Missing Value Ratio, and Random Forest play a critical role in helping us identify the right variables we should consider for our prediction modeling.

Gradient Boosting Algorithms

Gradient Boosting algorithms combine multiple weak algorithms to create a more powerful algorithm. Oftentimes, you will get into scenarios where the data set could be massive or there could be several distinct characteristics present in the data set. In those scenarios, you could use an ensemble of learning algorithms that combines the predictive power of several base algorithms to improve robustness.

Deep Learning Primer for Managers

Deep learning is based on the principles of connectionist systems—a collection of connected units or nodes called *artificial neurons* (a simplified version of biological neurons) transmitting signals to one another. The artificial neuron that receives the input signal processes it and then produces an output signal that is propagated to other artificial neurons it is connected to.

In the case of neural network implementations, the artificial neurons and connections typically have a weight that gets adjusted during the learning process by applying some kind of a derivative function (also known as an activation function). The weight increases the signal at a connection if it contributes to prediction accuracy. Conversely, the weight decreases the signal at a connection if it does not contribute to the prediction accuracy. The artificial neurons are governed by a threshold value so that only if the aggregate value of the signal crosses the threshold is the signal propagated further. Typically, artificial neurons are organized in layers. Signals travel from the first (input) layer, through the subsequent layers, finally arriving at the last (output) layer. A neuron weighted more heavily than another exerts more influence on the next layer of neurons. The final layer sums together these weighted inputs to come up with the final answer.

The math is fairly simple: $f(x^n) = y^m$

Where f is the neural network, x is an n dimensional vector of input variables, and y is an m dimensional vector of output variables. Notice though that n can be different than m, and that is very important.

For example, the input could be millions of pixels, while the output is compressed to a few numbers with meaning, e.g., the probability the image is an image of car/house and so on.

Now that we know roughly what an ANN (Artificial Neural Network) is, let's look at one (see Figure 4-1).

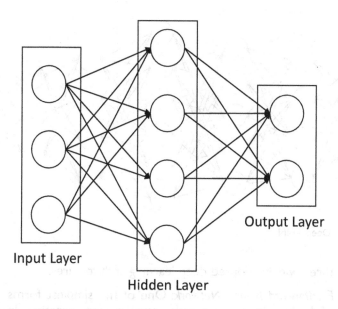

Figure 4-1. An artificial neural network (ANN)

Figure 4-1 shows a shallow (not deep) neural network where *n* is 3, *m* is 2, and we have a hidden layer.

The white circles are *neurons* and neuron connection (the gray arrows) are *synapses*. We usually call synapses weights in terms of ANN.

Weight in this context is just a number. Information flows from left to right, therefore you take a value on the left and move it to the right by multiplying the input value by weight. In other words, the value of a neuron in the hidden layer is the sum of all values from the previous layer multiplied by the weight of the connection. The neuron then holds the final value.

The *goal* is to find the weights that will produce a meaningful output for us.

Learning is the process of finding these weights. One thing you need to realize is there are a lot of weights. In this example, we have nine neurons in total, yet there are 20 weights. It gets even worse as you add more neurons. "Guessing" these numbers usually takes a long time and requires quite a lot of resources.

Guessing is a sophisticated approach of finding the values that will work the best.

Deep ANN is simply a ANN with more than one hidden layer, as shown in Figure 4-2.

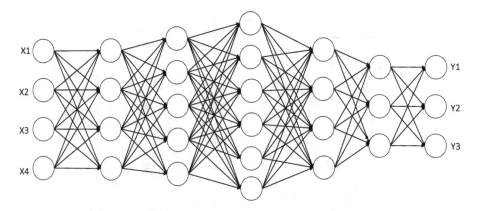

Figure 4-2. Deep ANN

There are three widely adopted deep learning architectures.

- *Feedforward Neural Network:* One of the simplest forms of ANN, where the signals travel in one direction. In simple words, the learning is achieved through the front propagated wave.

- *Convolutional Neural Network:* Similar to feedforward neural networks, but specialized to recognize images through learnable weights and managing biases and convolutions (recognizing edges of an object in the image).

- *Recurrent Neural Network:* In contrast to the previous two neural networks, RNN is designed to recognize sequences.

  The RNN learning process starts once the first layer computes the product of the sum of the weights and the features. The output then follows the feedforward neural network principle. From one layer to the next, each neuron remembers some information it had in the previous step. This makes each neuron act like a memory cell. Remembering what information the neurons may need for later use during the learning process is extremely important. While the front propagation wave will keep doing the incremental learning as it passes on signals to each successive layer in the network, the backpropagation needs to keep validating the prediction and use the learning rate or error correction to make changes to the weights so that the network gradually moves toward making the right prediction.

The three most common ways that data scientists use deep learning to perform object classification are:

- *Training from scratch:* You gather a very large labeled data set and design a deep neural network architecture that will automatically learn the features from your data and create the model for you.

- *Transfer learning:* Instead of building your own deep neural network architecture, you leverage deep learning architectures created by others (such as AlexNet or GoogLeNet) and use the transfer learning approach to your data. Since you are starting with an existing network, you have a head start. After making some tweaks to the existing network, you can train your network to perform newer tasks. Data scientists prefer transfer learning because the architecture is pre-built and tested on millions of data points. This way you can solve your problem with less data and it does not become computationally intensive.

- *Feature extraction:* Deep learning can be effectively used as a feature extractor as well. Since all the layers in the network are tasked to learn certain features from input data, you can possibly pull out these features from the network at any time during the training process and then use them as inputs to traditional machine learning models.

Deep Learning Algorithms and Architectures

Now that we have a high-level understanding of what deep learning means and how neural networks help in extracting meaningful outcomes from data, it's time to discuss some of the popular deep learning algorithms and architectures.

Feedforward Neural Networks

A feedforward neural network has the following construction:

- One or more layers, including a hidden layer.

- Each layer has a number of neurons in it.

- Each connection between two neurons has a weight.

- Each neuron of a layer is typically connected to *every* neuron of the previous layer (you can turn them off by setting their weight to 0).

- The learning process starts by applying an activation function on the neurons in the input layer. The resulting values are then propagated forward to the next layer's units using the weighted sum transfer function for each unit.

- The result of the output layer is the output of the network.

How do you measure the accuracy and the error?

The network error is minimized by applying gradient descent optimization algorithm. *Gradient descent* is used to find the minimum of a function. The optimal value is that at which the error achieves a *global minimum*.

To understand gradient descent, imagine the path of a river originating from the top of a mountain. The goal of the river is to reach the bottom-most point flowing down from the mountain. Now, if there are no deep pits in between, the river won't stop anywhere before arriving at its final destination. This is the ideal situation we desire. In machine learning context, we say that we have achieved the global minima of the solution starting from the initial point.

However, if there is a deep pit in the path of the river, then the river gets trapped and does not achieve the global minima. In machine learning terms, such a pit is referred to as local minima, which is not desirable.

Learning Rate Decay

Learning rate decay is an important performance tuning parameter. In the initial phases of training, you can afford to making large changes to the weights. However, as the time progresses and you add more iterations, it is desirable to have a decreasing learning rate decay so that with incremental smaller training updates, you can arrive at learning good weights and get closer to the desired final states.

Backpropagation

Backpropagation provides a mechanism to adjust each weight between two neurons by reflecting on the output error. The errors are first calculated by observing the output units (based on the difference between the target and predicted values), and then propagated back through the network to update the weights applied during the previous layer. The final goal is to attain a global minimum.

Dropout

There are chances of overfitting your learner if you have a large number of parameters for your deep neural net. Dropout is a technique to address this problem.

The idea is to randomly drop units (along with their connections) from the neural network during the learning process. This prevents units from adjusting too much. During the training process, you can configure to dropout samples so that the learner approximates the effect of averaging the predictions and reduces chances of overfitting.

Batch Normalization

Deep neural networks, at the beginning of the learning process, assume some initialization of weights, which are far away from the final state of the learned weights. During each iteration, these weights undergo careful tuning to reduce the errors. During this iterative process, it is possible that a small change in the initial layers may inadvertently lead to a large change in the later layers. Hence, it is prudent to keep the learning rates small, such that only a small portion of gradients corrects the weights. Batch normalization helps regulate the weight corrections through a series of min-batches, accelerating the learning process.

Vanishing Gradients

A human brain stores an abstract representation of the real world somewhere deep inside. Similarly, the hidden layers in a deep neural network also store abstract representations of the training data. The more hidden layers there are in a deep neural network, the more abstractions (more enrichment) there are on top of the previous hidden layers. However, if we keep increasing the number of hidden layers in a deep neural network, we get to a situation where the usefulness of backpropagation becomes meaningless, since there is no more gradient descent improvement we can achieve. In effect, as less learning information is passed back, the gradients begin to vanish and eventually become inconsequential to adjust the weight of the network.

Autoencoders

The network's goal is to "recreate" the input by learning the core features that define the data at a conceptual level so that it can reproduce the data by following a compact representation technique.

The network, instead of learning "mapping" between the input data and corresponding outputs, learns the internal structure and features of the data itself. How? In a neural network, the internal abstraction of data (most important, the features of data in reduced dimensions) resides in the hidden layers. The autoencoder exploits the hidden layers to develop its learning process.

Boltzmann Machines

Boltzmann machines are *neural networks that apply probability distribution principles the learning mechanism.* The learning process consists of two phases (positive phase and negative phase) working in tandem.

During the learning process, the network gains a perception of how the output data was created from the input data (you can call it the positive phase). It tries to use the same perception to recreate the input data (you can call it the negative phase). If the generated data is not close enough to reality, the network makes adjustments to its perception and tries again. The main goal is to keep refining the perception so that the generated data becomes as good as the original input data.

Filters and Max-Pooling

Convolutional neural networks (CNNs) are a class of feedforward networks specialized to do image recognition.

CNNs apply image filters and max-pooling effects to learn from data. An image *filter* is a rectangle with associated weights.

Convolutional layers apply a number of such *filters* to the input image to create many subsamples of the original image. The result of one filter applied to the image is called a *feature map.* For example, if the input consists of a 32x32 image and the image filter has a subsampling parameter of 2x2, the output value would be a 16x16 image, which means that four pixels (each 2x2 square) of the input image are combined into a single output pixel. There are multiple ways to subsample, but the most popular method is max-pooling.

Max-pooling is a method to partition the input image into a set of non-overlapping rectangles, reducing its dimensionality and allowing for assumptions to be made about features contained in the sub-regions.

GAN (Generative Adversarial Network)

GAN is composed of two neural networks (generator and discriminator) competing with each other. The training process consists of the generator trying to produce data (by observing the patterns within the training data set) using some form of probability distribution. The discriminator acts like a judge, deciding if the data generated is closer to the distribution of the training set.

Long Short-Term Memory (LSTM)

We do a lot of prioritization all the time, when we get new information, we immediately judge the importance of the new information and then start doing the re-prioritization activities. Turns out that the conventional feedforward networks don't do this. When a new information is added, they transform the existing information completely by readjusting the weights. Because of this, the entire information is modified. On the whole, there is no consideration for "important" new information and "not so important" new information.

There are many real-world problems where remembering a sequence of events is important to arrive at the final predictions. LSTMs are a new breed of neural networks designed to selectively remember or forget things.

An LSTM network uses memory blocks called *cells*. The memory blocks are responsible for remembering information. Updates to the memory are done by applying different mechanisms called *gates*.

Forget Gate

A forget gate applies filters to the cell state to either remove past information that is no longer required for the LSTM in the learning process or to ignore new information that is of less importance in the learning process.

Input Gate

The input gate is responsible for adding information to the cell state. This is basically done in a three-step process.

1. Creating an input vector containing all possible values for the cell state.

2. Controlling which values need to be added to the cell state by invoking a sigmoid function as a filter.

3. Multiplying the value of the filter to the input vector and triggering an activation action to complete the input operation.

This three-step process ensures that only important and non-redundant information is added to the cell state.

Output Gate

The output gate also follows a three-step process:

1. Creating an output vector after applying an activation function to the cell state.

2. Controlling which values qualify to be considered as output by invoking a filter function.

3. Multiplying the value of the filter to the output vector and triggering an activation function to complete the output operation so that the output progresses to the next layer.

NLP Primer for Managers

Natural Language Processing (NLP) is a specialized field at the intersection of computer science, artificial intelligence, and computational linguistics.

By utilizing NLP, we can analyze text data and perform numerous tasks, such as automatic summarization, named entity recognition, relationship extraction, sentiment analysis, topic segmentation, etc.

At a broad level, NLP consists of two specific tasks:

- Natural Language Understanding
- Natural Language Generation

Natural Language Understanding

NLU is the process of understanding the meaning of a given text. It tries to get a holistic view of the nature and structure of the words in the text by resolving various ambiguity:

- *Lexical ambiguity:* Words have multiple meanings
- *Syntactic ambiguity:* Sentences have multiple parse trees
- *Semantic ambiguity:* Sentences have multiple meanings
- *Anaphoric ambiguity:* Phrases or words that were previously mentioned but have different meanings

Once the ambiguities are resolved, NLU attempts to derive the meaning of each word by using lexicons (vocabulary) and a set of grammatical rules to resolve issues where different words have similar meanings (synonyms) and where words have more than one meaning (polysemy).

Natural Language Generation

NLG is the process of automatically producing text in a readable format with meaningful phrases and sentences.

Natural language generation consists of three key tasks:

- *Text planning*: Ordering the content to give it a structure.

- *Sentence planning*: Sequencing words and sentences to represent the flow of text.

- *Realization*: Correcting the sentences grammatically to represent the desired meaning of the text.

Here are a few examples of how NLP is applied to text.

- *Sentence segmentation*: Identifies sentence boundaries, i.e. where one sentence ends and another sentence begins. Long sentences are often broken with punctuation marks like commas. For example, "I met a friend after a long time, we had coffee together." can be split into two sentences. Sentence 1 is "I met a friend after a long time" and Sentence 2 is "We had coffee together".

- *Tokenization*: Identifies different words, numbers, and other punctuation symbols within the text. For example, "My car keeps giving me alerts, as the doors are not properly closed." can be tokenized as [My] [car] [keeps] [giving] [me] [alerts] [,] [as] [the] [doors] [are][not] [properly] [closed][.]

- *Stemming*: Reduces the inflectional forms of each word into a common base or root. For example, "Give, Gave, Giving" can be reduced to the root word "Give".

- *Part of speech (POS) tagging*: Assigns each word in a sentence with a part-of-speech tag, such as noun or verb. For example, "If you build it they will come." can be POS tagged as:

  - IN: Prepositions and subordinating conjunctions: "If"

  - PRP: Personal pronoun: "you"

  - VBP: Verb: "build"

- PR: Personal pronoun: "they"

- MD: Modal Verbs: "will"

- VB: Verb base form: "come"

- *Parsing*: Involves dividing given text into different categories. For example, "Soum and Sachin went into a bar." can be parsed as (S(NP(NP Soum) and (NP(Sachin)) (VP(went (PP into (NP a bar))))

- *Named Entity Recognition*: Identifies entities such as people, location, and time within the text. For example, "Sachin is traveling to the US on Sunday." can be recognized as [person = Sachin], [location = US] and [Time = Sunday].

- *Co-Reference Resolution*: Finds all expressions that refer to the same entity. For example, "Sachin bought a phone. He thinks it is worth the money." can be coreferenced as [Sachin, He] and [Phone, it].

Table 4-1 provides a list of deep learning algorithms that achieve common NLP tasks.

Table 4-1. Deep Learning Algorithms and Common NLP Tasks

Deep Learning Algorithms	NLP Tasks
Feedforward Neural Network	• Part-of-speech tagging
	• Tokenization
	• Named entity recognition
	• Intent extraction
Recurrent Neural Networks (RNN)	• Machine translation
	• Question answering system
	• Image captioning
Recursive Neural Networks	• Parsing sentences
	• Sentiment analysis
	• Paraphrase detection
	• Relation classification
	• Object detection
Convolutional Neural Network (CNN)	• Sentence/text classification
	• Relation extraction and classification
	• Spam detection
	• Categorization of search queries
	• Semantic relation extraction

Text Preprocessing

Text is the most unstructured form of all the available data. Before we can apply NLP techniques to text data, we have to ensure that the text is preprocessed. How?

There are three key intermediation steps applied to text data to make it ready for analysis (see Figure 4-3).

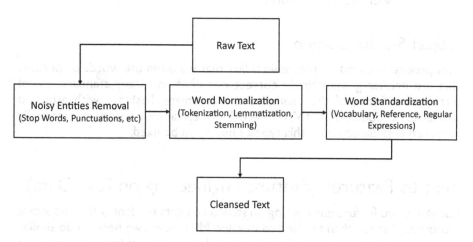

Figure 4-3. Illustrating the text preprocessing steps

Noise Removal

Any piece of text that is not relevant to the context of the sentence can be considered noise. Examples of noisy entities are language stop words (is, am, the, of, in, etc.), URLs or links, social media entities (mentions, hashtags), punctuation, etc.

One approach to noise removal is to a) tokenize the entire text, b) refer to a dictionary of noisy entities, c) eliminate those tokens present in the noise dictionary. Another approach is to use regular expressions while dealing with special patterns of noise.

Lexicon Normalization

The occurrence of multiple representations of the same word in a sentence can cause noise as well. For example, "eat," "eating," and "ate" all are different variations of the root word "eat". Although they demonstrate different meanings purely from a grammar perspective, the root word is the same. Lexicon normalization converts all the disparities of a word in a sentence into a normalized form (also known as a *lemma*).

The most common lexicon normalization practices are as follows:

- *Stemming*: Stemming is a rule-based approach to stripping the suffixes that we normally add to a root word to convey a specific meaning. Examples are "ing," "ly," "es," "s," etc.

- *Lemmatization*: Lemmatization is the process of reducing a word to its root form.

Object Standardization

This process is based on the general fact that we often use words or phrases that are neither grammatically correct nor confirm to any standard lexical dictionaries. Some of the examples are acronyms, hashtags with attached words, and colloquial slangs. With the help of regular expressions and manually prepared data dictionaries, this type of noise can be fixed.

Text to Features (Feature Engineering on Text Data)

Just like we do feature engineering on structured data to identify the important attributes that can then be used to develop ML models, we need to do similar activities on the preprocessed text data to determine text features. There are several techniques to construct useful text features, discussed next.

Syntactic parsing: Syntactical parsing analyzes the words in the sentence for grammar and their relationship among the other words.

Dependency trees: Sentences are composed of words following a particular order to convey the meaning. The relationship among the words in a sentence is determined by the dependency grammar. For example, consider the sentence: "Recommendations on BCCI operations and functioning were submitted by Lodha Committee." The relationship among the words can be observed in the form of a tree representation, as shown in Figure 4-4.

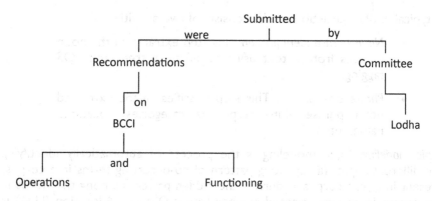

Figure 4-4. Illustrating the tree structure for sentence parsing

The tree shows that "submitted" is the root word of this sentence, and is linked by two sub-trees. Each sub-tree has another sub-tree with dependency and relations. The dependency tree helps identify text features.

Part of speech tagging: The POS tags help us understand the structure of the sentence. POS tagging is used for many important purposes in NLP, for example, in these two sentences:

- "Book my tickets."
- "This is a good book."

"Book" is used in a different context in both the sentences. In the first sentence the word "book" is a verb instructing to purchase tickets, while in the second sentence it is used as a noun and is talking about a tome that you read.

Entity Extraction (Entities as Features)

Entities are the most important objects in a sentence—noun phrases, verb phrases, or both.

Consider this sentence:

The order processing system is giving an error code of 1048.

Entity extracts would identify the following: order, error, 1048. This key information is more than enough to troubleshoot and provide resolution options.

A typical entity-extraction model consists of two activities:

- *Noun phrase identification:* This step extracts all the noun phrases from a text using dependency trees and POS tagging.

- *Phrase classification:* This step classifies all the extracted noun phrases into respective categories (locations, names, etc.).

Topic modeling: Topic modeling is the process of automatically identifying the different topics (a repeating pattern of co-occurring terms in a corpus) present in a text corpus. It derives the hidden patterns among the words in the corpus in an unsupervised manner. Latent Dirichlet Allocation (LDA) is the most popular type of modeling algorithm.

LDA algorithm: This algorithm assumes that documents are produced in the following fashion.

When writing this chapter, we:

- First conceptualized the key theme, "black box". Then we outlined what topic mixture should go into the chapter. For example, we decided on having sections covering overviews of ML, DL, and NLP and then covered the nature of AI as a black box.

- Then we added a few high-level restrictions, like the number of pages (people lose interest reading lengthy chapters).

- Then we went into the art of translating our thoughts to actually writing the words, sentences, paragraphs, and sections.

While this narration reflects how humans think when they want to write a document, a machine needs to understand this process inside and out.

The machine approach would be:

- First apply a tokenization approach to the entire document, then do a continuous bag of words to generate distinct topics appearing in the document.

- Then using the topic as a reference point, learn which words were used to generate the topic (the word "explainability" with 80% probability defines "black box" as the topic, "parsing" with 55% probability defines "NLP" as the topic, and so on).

N-Grams as Features: A combination of N words together are called N-Grams. N grams (N > 1) are generally more informative as compared to single words (unigrams) as features.

Skip-Gram

The main idea behind the skip-gram model is to determine whether two different words *are similar, if they share similar context.* For example, in the sentence "to err is human," if you use the term "mistake" or "screwup" instead of "err," the sentence is still a meaningful sentence and means the same thing (i.e., "is human").

Skip-gram applies a context window (a window containing k consecutive terms). You skip one of the words and try to train a neural network that takes all terms except the one skipped and predict the term you had skipped.

Continuous Bag Of Words

In the continuous bag of words model, the goal is to use the context surrounding a particular word and then predict that particular word. In a large text corpus, every time you see the word of interest, you consider the surrounding words (context words). You feed the context words to a neural network to predict the word in the center of the context.

Statistical features: Text data can also be quantified directly into numbers using several techniques.

Term Frequency-Inverse Document Frequency (TF-IDF): TF-IDF is primarily used for information retrieval purposes. The goal is to find the occurrence of words in the documents without the exact ordering.

For example, you have N documents. In any document D, TF and IDF are defined as:

- *Term Frequency (TF):* TF for a term "t" is the count of the term "t" in a document "D".

- *Inverse Document Frequency (IDF):* IDF for a term is the logarithm of the ratio of the total documents in the corpus and the number of documents containing the term "t".

- *TF.IDF:* TF.IDF gives the relative importance of the term "t" in the corpus.

Neural Networks as Black Boxes

Much like how our brain works, the inner workings of a neural network consist of layers of interconnected neuron-like nodes, an input layer, an output layer, and a number of intermediate "hidden" layers. The nodes themselves carry out mathematical operations based on the weights associated with the nodes and pass on internal representations of the input data to the subsequent layers, finally summarizing the weights at the output layer with the prediction. Deep neural nets have more than one hidden layer. By feeding thousands upon thousands of data to the neural network, you enable it to gradually fine-tune the weight of the individual neurons in a specific layer. The end result is a complex reading of all the neurons across all the layers weighing in to produce the final output. The journey from raw input to final outcome is so complex and goes through so many iterations and fine tunings that neural networks have gained the notoriety of being black boxes, in that they can't explain how they arrived at the final result. The only thing we can do is to try to create the best model possible and train it with as much unbiased data as we can get our hands on.

The key to the neural network training is the process called *back propagation*, in which intermediate-layer settings are progressively modified until the output layer arrives at an optimal match to the input layer. The more neuron-like nodes you have—the more weights and intermediate layers—the more complex the learning process is, even if you get to a fairly accurate prediction.

It's one thing to have a model that gives accurate predictions with previously unseen data, and it's another thing that you have no explainability of how your model arrived at the accurate prediction. There are many real-world situations where it is extremely desirable to have a view of the internal decision-making process in detail, because the consequences can be catastrophic.

Peering Inside the Black Box

The black boxes in airplanes are also known as flight data recorders. They record pretty much everything that happens during the flight journey. It is an extremely secure device and becomes a critical source of truth for researchers and investigators, if anything happens to the flight.

The black box (ML/DL algorithms) of AI has taken on the opposite meaning—the algorithm delivers predictions, but it doesn't necessarily tell you how it learned to predict and what it took into account while it was learning.

The developers of these "self-learning machines" also acknowledge that once a ML/DL algorithm is trained, it can be difficult to understand why it gives a particular response to a set of data inputs. This lack of transparency raises issues about the autonomy, decision-making, and responsibility aspects of AI.

There are two sets of activities that define how a machine goes about its learning activities:

- Knowledge gathering (through input data sets and that help validate and improve the prediction confidence levels) is used to arrive at accurate predictions.

- Embedding the predictive models into applications to not only automate decision-making but to also learn continuously and improve continuously.

Once the prediction system is well trained, it is time to put it into action. We see two prediction systems cluster into two types.

- Type A applications, where the consequences of prediction systems can have adverse effects on people's lives. These include medical diagnoses, loan processing applications, CV screening applications, self-driving cars, etc.

- Type B applications, where the consequences of prediction systems may at the most leave people dissatisfied. These include which movie to watch, which restaurant to dine at, which route for commuting will be faster, which news items you would like to read, which items you would like to buy, etc.

We are less concerned about the AI system's opaqueness when it comes to Type B applications, whereas we are extremely keen to understand what goes into making the predictions in the case of the Type A applications.

Neural nets learn by processing input data through multiple interconnected layers. The first layer serves as an input layer, where "neurons" receive raw inputs (such as pixels in a photograph of a human face). The neurons then apply certain weights and arrive at a score according to a mathematical rule. The aggregated score from this layer is then passed on to the next layer of nodes. This process continues across all the layers and all the nodes in the network until you get to the final output layer. The last layer of the network serves as the output layer and aggregates all of the scores in the previous layers into a prediction: "This is a picture of a human face," for example. The term "deep" means more number of layers; it can contain anywhere from three to hundreds of layers.

If the prediction is wrong, the neural net will then do a reverse tracing, tweaking the links between nodes, adjusting the mathematical rule, progressively steering the lessons closer to the desired result. By tuning the parameters to satisfy millions of examples, the neural net creates a model that can classify new images or perform actions under conditions it has never encountered before.

This learning process, known as deep learning, allows neural nets to create AI models that are specialized in solving a particular problem (such as image recognition, voice recognition, etc.), but are too complicated and too tedious to code by hand. What's commendable about deep learning is that you don't have to tell the system what to look for. You give it a few million pictures of cats, and the image recognition specialist will figure out what a cat looks like.

Because neural nets essentially program themselves, they often create an abstract representation of data that no human can fully understand. The decision-making process is encoded in the billions of back-and-forth signals between nodes.

Many experts find this non-explainability worrisome. It doesn't matter in a man versus machine game, but it seriously matters in the case of a driverless car. If the driverless car gets into a serious accident, it is simply not acceptable to say, "We just don't understand why the car did that."

The irony is that, we (humans) are building the prediction systems, and the wider the range of use cases we go after, the more avenues for the AI to successfully mimic humans, and hence the more adept the AI becomes in understanding why we do what we do, along with all of our biases, prejudices, and limitations.

Unmasking AI

Debates are raging to address the concerns surrounding the opaqueness of AI. A methodical approach to make AI explainable is still years away, but a number of approaches are in the labs. The approaches may vary but their goal is the same: to ensure that our machines do not evolve too far beyond our ability to understand them.

Some researchers are attempting to embed "explain training" modules into AI systems, so that any tinkering done with the inputs and resulting output can be documented automatically. This approach may work well in less complex AI algorithms; however, when you feed lots of data and your algorithm has lots of layers and then you have the "explain training" modules keep an eye on everything, the algorithm will inherently slow down the learning process, which is not a desirable effect. Others are attempting to develop "probe modules" that can do a dipstick analysis into the network's behavior and understand how these systems learn.

The "observer approach" treats an AI system like a black box. It is a pre-built neural net specialized to solve a particular task. Instead of feeding the entire data set, you experiment with it by feeding incremental data, piece by piece, and observing its behavior. For example, instead of feeding the complete picture of a car at once, you feed the neural net pieces of a car one at a time and observe which parts (headlights, front bumper, side view mirrors, windshield, partially visible steering, or something unexpected like a cracked windshield) lead the AI system to make a correct classification.

"Explanation net" is another approach to probing a neural net with a second neural net. The objective is to gain deeper insight into the tweaks and adjustments that happens in the original neural net.

While these approaches no doubt will bring some clarity as to how deep learning networks arrive at decisions, until we get to that state, we have very little choice but to develop an element of trust on AI delivering the outcomes, in the same way there is an inherent trust in humans who are responsible for decision making. For example, we accept the decision delivered by jury members and at best may contemplate which facts were instrumental in swaying their decision, but we do not go to the extent of quizzing them or scanning their brain activities to understand how they arrived at the final decision.

Training ML and Our Tolerance for Error

It doesn't take a tremendous amount of training to begin a job as a cashier at a gas station. Even on their first day, cashiers posses certain basic accounting knowledge to get started. On their first day, new cashiers may appear to be slow, inefficient, and might make more mistakes than their experienced peers. We generally have a good amount of tolerance and accept the fact that the cashier is going through a learning phase and it is just a matter of time that he learns the job.

We don't often think of it, but the same is true of medical practitioners. We take comfort in the fact that medicine as a practice is regulated by the government and requires not only passing a grueling curriculum but also internship experience of thousands of hours, before one gets a license to practice medicine. If someone wants to pursue a specialization such as heart surgery, requires an additional grueling curriculum and lots of internship experience to observe and learn from the experts. Once someone gets a license to practice medicine, it doesn't stop there. The medical practitioners continue to improve from on-the-job experience.

What constitutes "good enough" is based on our bias toward tolerance for error. Our tolerance is much higher for a new cashier, and much lower for doctors. We have different definitions for "good enough" when it comes to how much training humans require in different jobs.

The same is true of AI. Algorithms need to undergo training, just as doctors and cashiers do. They are also subjected to the "good enough" question and there are trade-offs to be made.

Let's take the example of the AI that manages your Gmail. The Gmail application applies spam-filtering to incoming emails, organizes your inbox according to your preferences, sets up your calendar entries based on the content in your mail, sends out automated responses, etc. But, it also fails quite frequently. We don't trash Gmail; we consider it "good enough" and through our continued indulgence with Gmail (correcting its mistakes), we provide a good training ground for the AI at our expense.

In contrast, we have low tolerance for error in driverless cars. We never had driverless cars before, there was no training data, so how did the AI learn? We took a clever approach of asking human drivers to take the first generation of autonomous cars, drive hundreds of thousands of miles, and trained the AI on the job. It was like a driving instructor taking a learner on supervised driving experiences before letting them drive on their own. The key question is, should we accept this AI behind autonomous vehicle as good enough?

In order to get better, the AI behind autonomous vehicles needs to learn in real situations, unsupervised. But putting the immature AI in real-world situations means giving the passengers a relatively "young and inexperienced" driver. Suddenly, the risks are magnified and our tolerance for error hits rock bottom. However, until and unless the immature AI is encouraged to operate in real-world situations, it will never learn the many nuances of successfully operating and thriving in the real world.

Herein lies the tricky trade-off. Putting AI in the real-world situations earlier accelerates learning, but could have grave consequences (challenging our tolerance levels). Putting products in real-world situations later slows down learning but allows for more time to improve the AI.

Conclusion

Why do we care so much about the "explainability" of algorithms? In many situations, the concerns that algorithms should be implemented with a great deal of caution are justifiably right. However, as we continue to find new applications for machine learning algorithms, we should not let this focus on algorithmic explainability blind us from a harsh truth about the world: humans are predictably irrational and no more explainable than the most opaque algorithms out there.

This is why the distinction between the two types of AI applications—those replacing rule-based and repetitive human tasks and those replacing human judgment-oriented tasks—is important. When we focus specifically on AI applications in which algorithms are replacing human judgment-oriented tasks, we tend to seek for "explainability" of AI more than with the former tasks.

Humans Can't Explain Their Own Actions

Humans have cognitive biases. We often aren't even aware of how biases enter into our thought processes. When you ask people to explain their decisions, they feel uncomfortable explaining why they acted the way they did. For the same problem statement and the same actions taken, two humans will come up with very different explanations as to why they acted the way they did.

Algorithms Are Predictably Rational

An algorithm will give you the same answer every time. Yes, when we build these algorithms, if we are feeding them our cognitive biases and data biases, they will confirm to biased outputs, consistently. Hence, when we start putting stress on "AI explainability" as a critical factor, we must ask whether our outcome is more valuable than understanding the process of getting to the outcome.

In the next chapter, we discuss how by combining robotic process automation and AI, companies can transform themselves by pursuing an intelligent automation journey.

References

1. https://www.smithsonianmag.com/smart-news/weve-put-worms-mind-lego-robot-body-180953399/?no-ist

2. https://hbr.org/2017/03/the-trade-off-every-ai-company-will-face

3. http://www.zdnet.com/article/inside-the-black-box-understanding-ai-decision-making/

4. https://hbr.org/2017/01/4-models-for-using-ai-to-make-decisions

5. https://futurism.com/ai-learn-mistakes-openai/

6. https://www.technologyreview.com/s/607955/inspecting-algorithms-for-bias/

7. https://www.cio.com.au/article/631997/how-transparent-your-ai-black-box-systems-better/

8. https://www.scientificamerican.com/article/demystifying-the-black-box-that-is-ai/

9. https://thenextweb.com/artificial-intelligence/2018/02/27/bye-bye-black-box-researchers-teach-ai-to-explain-itself

10. https://www.sentient.ai/blog/understanding-black-box-artificial-intelligence/

11. https://www.analyticsvidhya.com/blog/2017/01/ultimate-guide-to-understand-implement-natural-language-processing-codes-in-python/

Intelligent Process Automation = RPA + AI

Robotic process automation (RPA) and artificial intelligence (AI) have traditionally been viewed as separate and somewhat unequal worlds—individually driving significant efficiencies for organizations.

You might be hearing terms like intelligent automation, service delivery automation, and cognitive automation, and that these offerings have the ability to completely automate your enterprise operations. Let's step back and critically examine the marketing talks versus the reality.

In this chapter, we discuss several best practices to understand what RPA means and how, if you combine the AI capabilities with RPA, you will be able to deliver significant step-change transformations in your organization.

© Soumendra Mohanty, Sachin Vyas 2018
S. Mohanty and S. Vyas, *How to Compete in the Age of Artificial Intelligence*,
https://doi.org/10.1007/978-1-4842-3808-0_5

Robotic Process Automation (RPA)

RPA is a class of software "robots" that mimics exactly how humans operate (by logging into a system, entering data, executing workflows, etc.). They work with business applications, such as ERP and CRM systems and many other applications. Simply speaking, since these software robots replicate human activities, to a large extent the business applications work exactly as they always have before, without significant human involvement. The term "robot" signifies the software-driven capability to replace or enhance a human task. RPA primarily attempts to memorize the activities humans do at the "presentation layer" (the user interface) of business applications.

In addition to the automating process, these software robots are much cheaper compared to employing a human, and the software robots do not demand overtime allowances, as they can work 24×7 if need be. The benefits of software robots just from a cost optimization point of view is extremely appealing. Besides cost savings, RPA also delivers other important benefits, such as:

- *Accuracy and compliance*: The software robots will execute the business processes in exactly the same way every time, all the time, with consistent outcomes.

- *Improved responsiveness*: The software robots are generally faster than humans, can work all hours without any fatigue, and can even scale to take on more loads at a fraction of the cost

- *Agile and multi-skilled*: The software robots can do a multitude of tasks provided the inputs-outputs rules-based paradigm is well defined and measurable.

A condition to apply RPA to any processes is that, the input, rules to process the input, and the output, need to be clearly defined without any ambiguity or dependencies with external systems. When one single process has multiple interfaces across other systems and expects the inputs from other systems in a synchronized manner, all such dependencies need to be clearly articulated; otherwise, bottlenecks will make the RPA inefficient. The customer on-boarding process is a good example of such a scenario, since it involves a number of different steps and systems, all of which need to be clearly defined and mapped.

Robots today can open and read attachments in emails; they can move files from one folder to another; they can follow programmed rules like if/then/else; they can extract data from forms or input data into forms; they can direct that data into data stores like databases and files or push the data into integrated systems like ERPs, CRM, finance, HR systems, etc.; and much more.

Applications of RPAs include:

- *Finance and Accounting*: Vendor management, T&E, invoicing, exception handling, issue resolution, payment runs/cycles, payables, receivables, collections, monthly closing, reconciliations, and accounting, to name a few.

- *Banking*: Data validations, data migrations between applications, report creation, form filings, claim processing, originations, servicing of loans, and so on.

- *Capital markets*: AML & KYC processing, data remediation, management reporting, client reporting, customer and advisor on-boarding, license and registration processing, payments and sweeps, reconciliations, asset transfers, corporate action processing, etc.

- *Insurance*: Notifications to agents for renewals, credit checks, data entry for registration, updating client info in multiple systems of records, payment decisions, claims processing, daily bank reconciliations, etc.

The software robots will do exactly what you have tasked them to do each time, every time. This characteristic is both their greatest strength and their greatest weakness. It is a strength because the robot will carry out the process compliantly and accurately. It is a weakness because if there is any change in the business rules or inputs, or even scenarios that were not envisaged earlier, the software robot will have no clue how to handle that. For example, invoice processing is a voluminous activity. You can design a software robot to read the invoice forms, extract the key entities, do a validation with your CRM systems and payment processing systems, and then release the payment. This is a very simple process if all the invoices you get adhere to a single template. Suppose you are onboarding a new partner and their invoice form is different from those you had programmed earlier. Your software robot will not be able to process it and you will have to reprogram it to accommodate the change.

This inability to adapt to changing conditions highlights two areas of improvements for RPA, both of which can be supplemented by AI capabilities.

The first constraint is that, when the input data is unstructured and does not have a pre-defined format, such as a customer email, or documents where there is generally the same information available but in variable formats (such as invoices), then NLP capabilities are needed to extract the relevant information, which then can be fed into the software robot to carry out the tasks efficiently.

The second constraint is that software robots are good at applying rules-based reasoning to a small number of specific steps, but they can't use higher order decision-making. For example, if you have have a relatively small number of business rules to apply and approve a loan application, software robots can do the job very well. However, when the loan approval process becomes complex with not only multi-branching multi-dependent rules but also requires cross-referencing with real-time credit scoring, you are staring at a judgment-related task and AI supplement.

Artificial Intelligence

In the previous chapters we discussed AI at length. In the context of RPA, let's revisit AI's capabilities in two different categories.

The Ability to Capture and Process Any Type of Data

The first AI capability is the ability to deal with any type of data. With Big Data technologies, we were no longer constrained to deal with structured data only. The schema-on-read flexibility enabled by Big Data technologies enabled us to collect and store any type of data (relational structured data, unstructured data, semi-structured data, machine generated data, audio data, video data, and images). NoSQL databases also gave us the capability to process this data in a scalable, distributed, and efficient manner. While these advancements were initially designed to collect and manage data for web related problems (search, document management, cataloging, network of relations, etc.), a few other types of problem statements emerged that required out-of-the-box thinking. For example, how do we recognize images and interpret speech?

Turning Data Into Insights

The second AI capability churns all these different types of data to glean insights from it. This is where ML, DL, NLP, and neural nets come into play. NLP helps extract meaning from unstructured data and ML helps develop prediction capabilities from the data. Deep learning and neural nets learn from the data to solve problems like recognizing a face in a photo, interpret spoken words, and generate spoken words.

In the context of RPA, these AI capabilities can certainly work as a complementary technology. For example, NLP can standardize and extract relevant data from documents, emails, transcripts, and logs even if the they are in free-from, or even when the entities of interest are in different places in the document. Take the example of an invoice form. The date might be in the top-left corner sometimes, and other times in the top right, the bill of material may be represented as unit times quantity in some forms, whereas in others they may all appear as line items one by one and then totaled at the bottom. The descriptions might have been in elongated version in one form and in a codified version in another. The tax amount may be written above the total value or as a separate entry below. Relying on RPA capabilities alone would make handling this variability extremely challenging. Even if you added enough rules to take care of all these different scenarios, the resulting program will be highly complex and unmanageable. With NLP and ML capabilities, AI can easily solve these problems.

Conversational interfaces (or chatbots) can be used to facilitate interaction between humans and other systems. You type your asks in a natural language form, and AI picks it up from there to process your query and respond back with relevant answers. If you don't want to type, you can talk to the system and AI will use speech recognition to understand your query, interact with backend systems, and then talk back to you with the relevant information. Despite of all these advances in AI technology, we are still living in the world of narrow AI. Meaning an AI that does image recognition won't be able to generate text or the AI that does NLP processing very well won't be able to provide recommendations or next best offers. Hence, in the context of RPA, AI supplements need to be considered. By considering the entirety of the problem you are trying to solve, you will have to stitch together various AI capabilities.

RPA and AI are two very different technologies, yet they complement each other very well. For example, AI can help process a variety of unstructured data and build a rich knowledge base, which then can be utilized effectively by the RPA technology to automate tasks. As businesses adopt RPA beyond the low-hanging automation task type of problems, they should start looking at AI to bring in self-learning capabilities to the very same RPA realm and make their business processes intelligently automated.

Automation and Intelligence: Working Together

When you combine the automation capabilities of RPA and self-learning capabilities of AI, you are looking at convergence of technology solutions that can not only improve your operations and customer interactions but also may create new business models. Many companies are beginning to recognize the potential benefits of combining these two technologies to solve problems, especially when operations efficiencies are magnified because of large legacy systems or older business processes with multiple hops and complex integration patterns.

Let's first understand why there's a need for intelligent automation.

The Need for Faster Turnaround

In a highly competitive and fast-paced business environment, the most critical enterprise metrics measure how responsive the company is to its customers and markets. For that to happen, the customer facing and market facing applications need to be almost real time and the backend systems also need to be transformed to match the speed of the frontend. This dual transformation challenge needs to be carefully orchestrated because most backend systems have evolved over decades of hand-coded programs. Organizations are in need of solutions that can bring in these dual changes quickly.

The Demise of "Lift and Shift"

Enterprises are not looking for changes by adopting new technologies to do the same tasks the same old way, but with appealing user interfaces or elevating user experiences. They are looking for real change where systems become intelligent enough to adapt to changing business dynamics and changing customer preferences. A simple lift and shift approach to new technologies won't make the enterprise efficient, rather the real need is to look for disruptive and demonstrable business value beyond the cost savings or ease of doing operations.

Expectations Are Elevated

Decision makers do not want to settle with robust yet placid business systems. They expect business systems to be adaptive and responsive. They want their business systems not only do the job it is designed for but also to provide them additional insights proactively so that they can take more informed and better decisions.

The Journey of Intelligent Process Automation (IPA = RPA+AI)

Humans codified rules-based systems to automate processes and solve the problems of yesteryear. Humans also developed AI technologies to learn from observations and solve problems that hinged on judgment. What if you combine these two distinct problem-solving techniques? In essence, intelligent process automation leverages the learning and adaptive capabilities of humans and interjects those into the process redesign initiatives. When you take traditional rules-based automation designs and augment them with self-learning capabilities, you get IPA that not only replicates human judgment and skills but also over time learns to do the tasks even better.

Automation, when embedded with judgment or automation supported with the human in the loop, is redefining the way tasks are being allocated and performed in an enterprise. To begin with, IPA was primarily used in the manufacturing processes and later as bits and pieces in other functions. Now, it is increasingly becoming an integral part of various enterprise functions across the board.

Organizations are putting in place holistic architectures of IPA systems because of its potential to be the critical vehicle for effecting enterprise-wide transformation. The time to embark on the " IPA journey" has indeed arrived. To automate or not to automate is no longer a question. The primary stimulus for this new enthusiasm stems from the appreciable double-digit improvement in productivity across organizations that have implemented RPA. RPA tools are set to evolve further as they imbibe augmented intelligence capabilities to mirror human intelligence. And as they become smarter, we will also see them perform more intelligent and complex tasks such as planning, budgeting, analysis, and decisions making that were perceived to be within the preserve of humans only.

Before we go into what constitutes IPA, first let's debunk a few myths about automation!

- *RPA will improve productivity and reduce cost*: Yes, by automating business processes, one can surely improve productivity and reduce cost. However, the associated aspects of change-management need to be carefully evaluated. Often, it is seen that the gains in productivity and reduced cost are eclipsed in additional overhead that one needs to put in to manage the change.

- *Before you apply RPA, you need to improve the maturity of your processes*: This is not a prerequisite, rather it is a good to have wish. The only objective of RPA is to break down your processes into smaller chunks and find ways to automate them. That way either the end-to-end process can get completely or partially automated. Either way, you will introduce transformative capabilities into your processes and operations.

- *Given the maturity of vendor tools, you can deploy software robots in production quickly*: The answer is—it depends. If the process in question is deep within some backend systems, has not seen any significant human involvement in the past, and is more or less a standalone sub-process, then the answer is clearly yes. However, if the process has dependencies with other systems, has a workflow with multiple human validation points, and deals with certain data that is fairly complex and sensitive, then you need to be careful not to rush into production deployment. It is advisable to have a pilot solution for a lengthy period of time, evaluate the effectiveness and efficacy of the solution, plan a proper change-management program, and then take it to production deployment.

- *The key success criteria are the number of bots you have deployed*: Nowadays, it has become a trend to quote the number of bots you have in production as a proxy to articulate your organization's automation journey and maturity. In our view, it is not the number but the efficacy and efficiency that matters. At the one end, you may be reducing the FTE count by automating manual activities, but at the other end, you are increasing the number of machines in your workforce mix. These machines will need supervision, governance, and maintenance efforts as well. Hence, utilization per 'bot' is probably a better measure for understanding the efficacy of automation.

- *Why RPA and then cognitive? One can straight away move to cognitive*: RPA is a straightforward process of mapping input to output, whereas cognitive means you are bringing in learning from data to dynamically do your automation. This requires a certain level of data, process, and knowledge base maturity. If you have not simplified your processes, the resulting data will be fragmented and disconnected, which means for you to apply machine learning, you have to do the heavy lifting of managing and consolidating your data, which in turn means you will take longer to train your machine learning models to deliver the kind of automation you are looking for. Hence, it is advisable to start with RPA. It becomes a stepping-stone in your automation journey and prepares the organization for becoming an autonomous organization in the long run.

IPA is not just one tool, rather it consists of five core technology capabilities, all integrated to provide a platform on which you can build and deploy intelligent automation solutions:

- *Robotic process automation (RPA)*: Software that enables you to develop "robots" to automate tasks that have a clearly defined set of inputs, rules-based processing patterns, and measurable outputs. Just like a human operator has an ID to log into business systems, an RPA also would have a user ID. It will memorize what the human operator does in the business system and start mimicking the rules-based tasks such as accessing user interfaces, performing data entry or validations, creating documents and reports, and triggering the next set of actions.

- *Smart workflow:* A process management application that integrates tasks performed by humans and robots. The smart workflow system will orchestrate handoffs between different users, including between robots and human users, to execute tasks efficiently and to monitor the end-to-end process in real time.

- *Machine learning:* Algorithms play a vital role in identifying patterns in data, such as new scenarios or rarely occurring scenarios being observed in the process flow. These scenarios, if not handled through prediction technologies, will create huge bottlenecks within the process flows. There are other scenarios where human validation is required to move to the next set of process steps. In such scenarios, an ML algorithm can provide recommendations and next best actions to the human worker to move on with the process efficiently.

- *AI capabilities:* A set of specialized AI capabilities to deal with unstructured data, including documents, images, videos, and voice. Business processes often end up managing and creating a variety of data types. In many scenarios, inputs to business systems could be in the form of images, videos, and voice. AI capabilities are required to manage these data types so that relevant information is extracted and passed on to the smart workflow and the ML programs that carry out the tasks.

- *Cognitive agents:* Self-sustainable modular executables acts like a virtual worker (or "agent") capable of executing tasks, communicating with systems or humans or other agents, learning from data sets and the environment, and even making decisions based on "emotion detection." Cognitive agents can be passive, coming to the aid of other agents or humans only when they are asked to. They can be active, observing the overall health of the process and jumping into action when they see bottlenecks.

Now that we understand how different technologies coming together provides a very powerful platform, the next question is, how does an IPA platform perform in real-world?

Let's consider the loan approval process. A human loans processor will point and click to several systems like CRM, loan accounts application, risk profiling system, legal and compliance system, external credit bureau system, etc., to provide a "business as usual" service. With IPA, robots will pull all the information from these different systems and arrange them neatly for the human worker to apply any judgments, thus saving lots of point and click effort. The NLP programs will kick off to scan through past interactions (emails, voice calls, complaints, and preferences) with the loan applicant and interpret text-heavy communications. The ML algorithms will provide alerts based on similar profiles and past risks or provide next best actions as recommendations. Finally, the cognitive agents will initiate interactive communications with the loan applicant and provide real-time tracking of handoffs between systems and humans (see Figure 5-1).

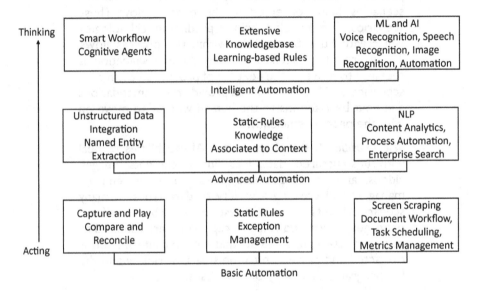

Figure 5-1. Smart workflows: from acting to thinking

Your organization needs to introduce automation and intelligence in an organized and thoughtful way. The following sections discuss a few common pitfalls that you need to steer clear of.

Create a Value-Based Strategy

Organizations should critically evaluate the hype behind many vendor tools. They need to establish a realistic view of the business benefits of intelligent process automation for their organization. They need to objectively assess the current status of their business processes, customer unmet needs, market trends, and employee productivity metrics as well as employee engagement index. Then they should identify a comprehensive set of opportunities that are value based, not just save money. They should do a proper study of how they can utilize internal and external data sets to transform both the frontend applications and backend systems.

Prioritize Opportunities

Given the potential of intelligent process automation to transform business, it is quite natural that you will get into situations where the business function leads will have their own set of priorities and the CIO organization will have their own set of priorities. One sensible approach to setting priorities across the organizational boundaries is to create a heatmap of intelligent automation opportunities across business processes. Taking a view of business value created against time to implement, you can come up with a list of business processes and then develop an IT readiness plan to portray your current capabilities versus what other technology changes are required to deliver the business transformations.

Identify Pilot Scenarios

When choosing opportunities, carefully consider the implications as well. Processes might be running on very old legacy code, no documentation available, multiple hand-offs across systems to complete the end-to-end process flow, and too many judgment calls embedded into the process flow. If these are the situations you uncover, it is pragmatic to step back and look for simpler systems or processes to deal with first. You need to choose a few simpler scenarios that can deliver immediate business value, and the success of these pilots will certainly help you make inroads into other complex scenarios.

Create a Roadmap

Too often we see that business functions and IT leaders jump into doing several pilots and then declare they are ready for large-scale enterprise-wide implementations. There is nothing wrong with doing pilots and taking smaller and simpler tasks into account. However, when you want to initiate enterprise-wide transformational projects, you will need a roadmap. The impact of intelligent automation goes beyond automating tasks. There are change-management impacts across functions and roles. Your roadmap should have a timetable and a clearly outlined set of line items to address the underlying changes to technology, organization, people, and operations. Top-down sponsorship is the key to a successful intelligent automation implementation. Refer to Figure 5-2 for an intelligent automation journey map.

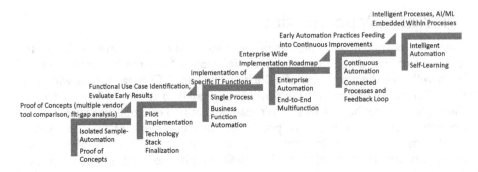

Figure 5-2. Intelligent automation journey map

Integrate Processes and Technology

There are enough low-hanging use cases (at least in the operations space) that can be successfully implemented using simple RPA; however, when you attempt to bring transformational power to complex processes cutting across business functions and IT systems, simple RPA tools fail to rise to the occasion. Vendors vary widely in their ability to support a holistic architecture that covers the five key components of intelligent automation platform we discussed earlier. Organizations need to critically assess vendor capabilities in this area, especially to ensure the new systems integrate seamlessly with the overall enterprise ecosystem. They must establish a performance monitoring mechanism to manage and monitor the overall intelligent automation projects.

Establish Operating Procedures and Governance Models

With automation mimicking human repetitive tasks and AI providing human-judgment types of capabilities, the enterprise workforce and standard operating procedures are expected to go through a sea change. As humans and cognitive agents interact in an environment of continuous learning, functional and technical teams must find a way to collaborate and understand who is responsible for what.

Historically, enterprises have evolved and have become better at managing the human workforce. Standards, best practices, and HR processes have been designed to assess and reward human performance. How would you execute these practices when you have machines in the mix?

More broadly speaking, companies may be steadfastly adopting the "automation and intelligence first" approach without paying significant attention to operating procedures and governance models.

Establish Change-Management Programs

Intelligent automation will change the way we have worked so far. For sure, many traditional jobs will lose its relevance and understandably there will be resistance from employees. Change-management programs are critical to address several challenges:

- Managers need to be trained to oversee processes in which robots and human work together.

- Employees need to be trained to adopt Agile methods, especially when they have to match the speed of machines.

- Executives need to be trained to establish clear guidelines around segregation of responsibility between human and machines.

- The company at an overall level needs to establish new communication channels to manage the employees' concerns around job loss and launch new programs to introduce new jobs (EQ skills, new technology skills).

Intelligent process automation is a gift and a riddle. It's a gift because it ensures fast payback if properly implemented, and it's a riddle because there are no such guidelines or precedents of how to successfully manage machines and humans that are working in tandem in the same environment. Next, we have made an attempt to outline how you would approach your intelligent automation journey. These are some of the best practices to help you define your own journey.

Your IPA journey: The First 100 Days

Days 1 to 30: Raise awareness, align functions, and mobilize, including strong top-down sponsorship: The first 30 days is all about engaging with your business executives and IT leaders and aligning expectations. Success of IPA initiatives is less about robustness of the technology solution. It is more about establishing a strong change-management plan and raising awareness across business and IT functions of how IPA will bring efficiencies into the organization. You must have a master plan that details the initiatives, the timelines, the extent of disruption or impacts expected, the change-management plans against each initiative, who is in charge of each initiatives, and so on. If your IPA roadmap is organized around the complete end-to-end process (starting from loan origination to final loan disbursement), instead of being organized around tasks (for example, a loan approval task), it will give you a holistic picture of how much automation you can achieve, where you still need human involvement, and what business outcome you will be delivering.

Days 30 to 60: Assess opportunities and conduct proof of concepts: Every organization is under pressure to deliver strong margins and introduce efficiency levers into their operations. Hence, it won't be surprising to see every business function clamoring with lists of processes they want to automate. If you ask them to prioritize, they might vociferously declare that each one of their initiatives is a top priority. Hence, you need to quickly decide on a methodology (from assessment to deployment) that will work in your organization. You need to create an objective-based framework for assessing and prioritizing automation opportunities. This will not only help you manage the mad rush for automation effectively, but will also help you identify quick-wins that can deliver business value early in your IPA journey. Beyond the high-level process automation maps, you must go down a few steps deeper to conduct proof of concepts to assess the complexity and usefulness of automation for a few of the identified automation opportunities. This will help you get a feel of the tentative timelines and any other technology you need or functional gaps you have. Most importantly the proof of concepts will give you a good enough opportunity to socialize the findings with business executives and IT leaders about the progress of your IPA journey.

Days 60 to 100: Establish governance model and begin development at scale: Once you are through with your proof of concepts and have a good enough sense of which product, tools, and processes you will go with, you need to fire on all cylinders to progress in your automation journey. At any point in time, you will probably have different processes at different lifecycles. Some will be at the requirement stage, some at the development stage, and some at the change-management stage. It is critical to have a governance model in place to address dependencies, bottlenecks, escalations, course corrections, etc. Infrastructure requirements will become a critical element to move into production at scale, beyond the first 100-day period. You need to define and start building an IPA CoE that will provide critical guidance around infrastructure planning, standards, best practices, technical and functional reviews, etc.

Your IPA journey does end with deploying bots into production. In fact, the journey starts after deployment. You need to track and monitor the effectiveness of your bots are determine whether they are delivering the business objectives. The success of the first few deployments needs to be spread across your organization so that the rest of the fence-sitters will get motivated to be part of the IPA journey. You will need "automation evangelists" to champion the IPA objectives.

Now, if your CEO is asking you to come up with an operating model to put the company on the IPA journey, keep in mind that the operating model needs to be anchored around a few key elements that will allow the successful completion of tasks by the key stakeholders. These elements are critical in designing a flexible, yet robust IA program.

15 Key Essentials For a Successful IPA Journey

While RPA may sound simple and easy to implement, and AI has its own technological complexities, the intelligent automation journey does require thorough planning, coordinated action, and lots of rigor to be successful. Organizations may do well to follow these key ground rules:

1. *Start with a proof of concept*: While RPA as a concept may be understood, showcasing RPA through a quick workable pilot project will trigger enthusiasm and help ward off cynicism.

2. *Set the right expectations*: Promise less, deliver more. This age-old wisdom holds true here too. It helps to set realistic expectations around the potential benefits. So, avoid creating hype or euphoria around what RPA can achieve for the business or individuals. Let people be impressed with the end results.

3. *Have a robust solution focus*: Invest efforts in building the right solution to address majority of the variations, handoffs, and process goals. Usually, 30% or less time is spent on actual BOT configuration.

4. *Identify and bring evangelizers on board*: Change is often resisted even if it is for common and individual good. Thus RPA-led transformation requires strong sponsorship and support. It is also important to identify functional leaders and opinion makers right at the beginning of this journey to implement RPA adoption seamlessly.

5. *Leverage complementing tools*: Be on alert to identify and explore tools that can complement and strengthen (e.g., OCR) RPA implementation.

6. *Follow the quick win delivery methodology*: Chances of success increase with smaller manageable sets of automations than large and complex ones. Follow an iterative process of fine-tuning the solution through build-test-deploy procedure.

7. *Choose processes wisely*: The success of first steps will have a significant bearing on the outcome of this migratory journey. So, choosing the first set of processes is very critical for the entire program to succeed. Initial processes have to be the ones where success is almost built-in while inconvenience or the pain of transition is minimal. Once the initial phase meets success, there will certainly be greater organizational support.

8. *Make IT an integral part of the journey:* Value from RPA has to be co-created by the business and IT teams. Given the short bursts of implementation and launch of automations, it requires the business and IT to work in harmony, in the absence of which time lapse and resource waste are bound to happen.

9. *Track and reap benefits simultaneously:* Actual benefits flow in only on productive redeployment of the saved hours. It is important to track actual savings on the baseline.

10. *Plan for sustainability:* Institutionalize structure and governance to productively manage automations that have been delivered and prioritize and deliver on the pipeline of opportunities. Adopt a holistic strategy to build, re-build, and sustain talent.

11. *Operating model:* Guidelines and policies required to execute tasks across the intelligent automation lifecycle; existing enterprise guidelines and policies needs to be updated to reflect the human-machine collaboration model, such as business continuity plan, data security and compliance, and release management.

12. *Standards:* Standards and practices to consistently manage intelligent automation activities, such as exception management and logging, coding standards, documentation standards, and deployment and readiness checklist.

13. *KPIs and metrics:* Measures to cover different aspects of the intelligent automation solution, such as process performance metrics and productivity metrics.

14. *Methods:* Frameworks/approach to ensure a consistent approach across functions, such as process evaluation framework, process prioritization framework, business case framework, change request framework, and exception management.

15. *Governance:* Clear segregation of responsibilities and review mechanism to ensure conflicts are swiftly dealt with, such as decision matrix/RACI framework, and meeting and reporting cadence.

Conclusion

There is certainly plenty of excitement around automation right now. RPA alone can deliver significant benefits; however, to make automation sustainable and to do real enterprise-wide transformation, AI capabilities need to brought into the mix. Product vendors have latched onto the buzzwords and have been marketing cognitive capabilities in their RPA tools in a big way. As you start your automation journey, it is therefore important to understand the reality versus the hype and plot your own roadmap.

On the cautionary side, RPA and AI are not cure-all technologies. However, by objectively assessing your own use cases and designing your automation approach by incorporating RPA and AI technologies into the processes, you can increase productivity and make your organization responsive.

Beyond the technology underpinnings, a successful intelligent automation implementation requires strong change-management programs to address uncertainties and impact on organizational and operational functions as well as on employees, who often are the receiving end of any large-scale transformational programs.

In the next chapter, we discuss the interesting world of cybersecurity, including the implications of AI on security, and how companies can safeguard their systems and processes.

References

1. https://disruptionhub.com/robotic-process-automation-artificial-intelligence/

2. https://fsinsights.ey.com/big-issues/Digital-and-connectivity/shaping-insurance-robotics-rpa-and-ai.html

3. https://home.kpmg.com/content/dam/kpmg/my/pdf/accelerating-automation-plan-your-faster-smoother-journey

Cybersecurity and AI

For a long time, cybersecurity was a war among humans. The protection mechanism was largely a "seal the borders" approach, via firewalls, proxies, antivirus software, access controls, dynamic passwords, and so on. However, these old methods are beginning to seem inadequate as the battlefield has changed to human versus machine. With the increasing number of entry points into the enterprise systems landscape and more connected devices as endpoints, news about security breaches are appearing more frequently than before.

Cloudification, IoT, and BYOD (bring your own device to work) are all giving rise to microenvironments that contain a lot of sensitive data about the user. If these devices fall into the wrong hands, this could certainly lead to grave consequences. The situation is further alarming when you consider the newer type of attacks, which are mostly machine engineered. For example, adaptive malware is created using machine learning techniques. This type of malware can infiltrate a system, collect and transmit data about that system, and remain undetected for days.

Applying the old approach of gathering information about data breaches, malware types, and phishing activities and then creating signatures is no longer potent enough to handle the next generation of cyberattacks. The threat is real and also magnified, primarily due to increasing digitization all around us. Standard automated threat-detection systems, although improving in sophistication, find it challenging to react quickly to unanticipated or newly formed threats.

© Soumendra Mohanty, Sachin Vyas 2018
S. Mohanty and S. Vyas, *How to Compete in the Age of Artificial Intelligence*,
https://doi.org/10.1007/978-1-4842-3808-0_6

A new approach is required to continuously monitor the large number of factors and detect what constitutes abnormal activity. In short, they need to apply self-learning techniques to spot what could be a malicious activity without being told what to look for! This new approach could be similar to our body's immune system, where the white cells and antibodies are continuously scanning and neutralizing any organism that does not fit the normal functioning patterns within the body. This is where AI comes into play. Machine learning algorithms can recognize potential security breaches or attacks by continuously observing what is an abnormal behavior, and if given the authority and the right credentials, they can automatically shut down systems under perceived threat, thereby reducing or isolating risks to the entire enterprise.

There are a few challenges with this approach though. Machine learning lacks the general knowledge required to distinguish real threats. This leads to too many false positives, too many false alarms, and frequent shutting down of systems. For AI to play a decisive role in cybersecurity, it must be treated with caution and applied to problems it can solve.

A hybrid human-machine collaborative approach to cybersecurity could be a potential solution. Neural networks can learn from enormous amount of data, including accessing publicly available cybersecurity literature to raise alerts that can be correctly classified by human cybersecurity experts. This human-machine collaborative approach would eventually make the machine smart enough to autonomously detect, classify, and take corrective actions. There are several examples of human-machine platforms in various stages of maturity.

AI2—from MIT's Computer Science and Artificial Intelligence Lab (CSAIL)—is a cybersecurity platform that uses machine learning combined with human experts. While machine learning, algorithms are trained on enormous amount of data to predict what could be potential threats. Human experts handle the judgment related tasks of validating and classifying the threats and associating severity tags.

Today's cybersecurity attacks conceal their presence in the systems thus making it a difficult task for the security analysts to pinpoint these attacks within a massive amount of system and network generated data. IBM's Watson uses NLP capabilities to analyze millions of documents, logs, and research papers that have never before been accessible to provide a more complete picture of the threat. Watson also generates real-time reports on these threats to greatly speed up issue identification and subsequent issue-resolution processes.

Microsoft recently acquired Hexadite to offer a fully automated incident response solution that enables organizations to investigate every cyberalert they receive and close out incidents in real time. Google attempting to develop cybersecurity offerings based on machine-learning capabilities for Android security.

Machine learning is capable of locating issues that might easily be overlooked in overwhelmingly large data streams. In order to understand how we can apply machine learning to cybersecurity, we need to first answer the question: what is our goal?

Broadly speaking, our first and foremost intent is to detect anomalies—to find malicious behavior or malicious entities. However, to find anomalies, we must first define what is not an anomaly! This is where the challenge lies. For example, starting from reading the morning news online to shopping to travel booking to carrying out work-related activities, we use our laptops in many different ways. There could be also infrequent patterns like downloading a game or organizing pictures from a vacation. How would you differentiate what is normal? It is highly possible that the game you downloaded may trigger some malware? Or the pictures you organized using freeware could actually introduce a security vulnerability? In essence, the most potent security threats are not just statistical outliers.

Applying Machine Learning to Security

Machine learning can introduce enormous capabilities and revolutionize the way cybersecurity has been handled to date.

- *Detection*: Track anomalous behavior amidst massive systems and network generated data in real time to raise alerts and assist humans in validating and taking appropriate actions.

- *Protection*: Identify and prioritize vulnerabilities to take action, including in some cases (where the prediction capabilities have reached significant confidence levels) addressing the vulnerabilities automatically.

- *Prediction*: Algorithms and humans working in tandem to significantly improve the machine's learning capabilities from each attack and developing even more sophisticated anti-malware techniques.

- *Termination*: With improved prediction capabilities, you can not only identify malware but also terminate it in an autonomous manner.

Now let's deep dive into applying different machine learning techniques into the cybersecurity related scenarios. There are two ways to look into the possible use cases:

- Apply supervised learning to the massive amount of good historical data (where the data set is labeled) to continuously improve prediction capabilities.

- Apply unsupervised learning to make some sense out of the massive amount of system and network generated data through clustering and dimensionality reduction techniques.

Supervised Machine Learning for Cybersecurity

The most talked about use cases of supervised learning in the context of cybersecurity are malware classification and spam detection. Classifying malware starts with identifying whether a file is harmful. Similarly, by observing patterns within the content, machine learning can categorize it as spam or not.

Supervised learning in cybersecurity has worked so well because of the availability of millions of labeled data. The neural nets get trained on the labeled data to identify abnormal patterns like intrusion detection, malware detection, phishing attacks, etc. In Gmail, for example, the supervised machine learning algorithm scans countless variables such as the originating IP address, originating location data, word choice, phrases in the email content, and a host of other factors to determine whether the email confirms to an abnormal pattern and then pushes the email out of your inbox folder and into the junk folder if it does.

With enterprises seeking solutions to protect threats emanating from the growing number of bring-your-own and choose-your-own mobile devices, supervised machine learning is being embedded on mobile devices. For example, Google is using machine learning to analyze threats against mobile endpoints.

The real benefit of supervised machine learning in the case of cybersecurity is that it can automate repetitive tasks like tactical firefighting to manually respond to tickets and run scripts to fix malware. This would free up analysts to focus on more important work like triaging threat intelligence.

For supervised learning to become much more effective, it requires massive amounts of good labeled data sets. In the absence of that, we cannot train our algorithms.

Unsupervised Machine Learning for Cybersecurity

Unsupervised learning doesn't have a goal to achieve; however, it can help reduce the clutter in massive data sets and make it comprehensible for further analysis by security experts. Context and expert knowledge base are two critical aspects to make sense out of raw data. This is where unsupervised learning can be of immense help.

For example, MIT's Computer Science and Artificial Intelligence Lab (CSAIL) developed an adaptive machine learning security platform that helps analysts review millions of logins each day and then contextualize the activities so as to reduce false positives to an almost negligible amount.

To better understand the role of the entities featured in disparate data sets, we need to not only extract the entities but also associate context to them. Rather than looking at network traffic logs in isolation, we need to add context to make sense of the data, such as information about devices, applications, or users, location specific data, whether the device is supposed to respond to DNS queries, and so on. If it is a DNS server, then this is absolutely a normal behavior, but if it isn't, the behavior could be a sign of an attack.

An Expert Knowledge Base helps human experts to quickly analyze the issues and find potential solutions. Machine learning algorithms like Bayesian belief networks, when applied to massive data sets, can result in graphical models and correlation factors that reason about uncertainty. This approach is very effective in building expert systems that do not necessarily solve all the problems right away, but can help make security analysts more effective by assisting them in their day-to-day activities. Instead of having analysts look at huge data sets and then cross-reference them with known issues and resolutions, which is a time consuming affair, the visual representations with correlation factors, dependency paths, and probabilistic measures provide a deeper understanding of the entities in a very short amount of time.

The CISO's Playbook for AI and Cybersecurity

Cybersecurity awareness is growing as more and more businesses are becoming digital. They are learning that their networks are vulnerable to attacks. The emerging consensus is that the IT department alone cannot handle security; all employees, especially C-level executives, have a part to play. In fact, executives are responsible for not only the security of data in their departments but also of broad employee adoption of cybersecurity best practices.

The risks vary by industry. For an electrical utility, the top cybersecurity priority is keeping the lights on by preventing cyberattackers from disabling the power grid. For a retailer, it's protecting credit card records and customer data. For the aviation industry, it's protecting the air traffic control system for the safety of passengers and flight crews. For the CISO (Chief Information Security Officer), cybersecurity is not simply about making a checklist of requirements, rather it is actively managing cyber risks at an acceptable level. If you are a CISO, you need to develop a cyber defense plan that is relevant to your industry.

Next are a few best practices to take into account when you are preparing your cyber defense plan.

Before a Breach

Preparing for a breach should be part of the daily security routine of any organization. Multiple layers of network security minimize gaps in protection. In addition to firewalls, intrusion prevention, and antivirus technology, you should also consider deploying advanced intelligence capabilities that inform you of emerging threats beyond your own network perimeter. Persistent security measures will also help protect off-premise and remote devices, such as cell phones and tablets.

Because the cyber threat landscape is always changing and new security solutions come to market all the time, you need to review your security strategy regularly. Questions that would help you formulate your strategy are:

- What is the current level and business impact of cyber risks to your company?

- What are your plans to address those risks?

- Do you have an escalation matrix and communication plan that identify executive leadership roles to determine the business impact of cyber risks to your company?

- Where does your cybersecurity readiness stand in comparison to industry standards and best practices?

- Do you have a systematic way of collecting, analyzing, detecting, and responding to cyber incidents on a regular basis?

- How comprehensive is your cyber incident response plan and how often is it tested?

Preparing for a cyberattack involves developing a coordinated response plan that considers every possible threat to your organization. Studying security best practices and failures of other companies, especially those in your same industry, is an excellent first step. In addition, discussions and workshops with counterparts in finance, human resources, marketing, etc. helps identify pertinent threats and develop defenses.

Like fire drills, you should develop scenarios and response plans that should be tested frequently and revised and improved as needed. You should develop mock scenarios of cyberattacks like, such as if website has been defaced, now the attackers are accessing confidential records, they are removing sensitive data from your network, etc. Your security analysts and AI systems must go through the process of discovering, identifying, prioritizing, and addressing the issues, as they would in a real attack. Beyond just developing an incident response plan, penetration and red team testing help establish the actual state of your security systems and procedures.

You need to perform assessments to thoroughly review your security environment and then create and execute realistic threat scenarios against your vulnerabilities that your security teams must then respond to.

Several cybersecurity assessments can determine how you can most effectively prepare for and respond to inevitable attacks:

- Compromise assessments apply extensive threat intelligence and security expertise to determine whether you have been breached in the past or are currently under attack. This assessment includes recommendations for further investigation, containment, and long-term security improvements.

- Proactive objective-based tests evaluate your security measures against the tools, tactics, and procedures used by attackers who typically target your industry.

- Penetration testing, red team operations, and other objective-based tests can detail risk, probability of exploitation, and potential business impact, and they provide actionable recommendations.

- Security program assessments review your security organization, practices, and procedures against the latest industry standards in critical security domains. This comprehensive assessment provides a roadmap with prioritized recommendations to close gaps based on vulnerabilities, attack trends, and any likely malicious activity.

- Response readiness assessments review your security operations and incident response capabilities for detecting and responding to attacks. The final deliverables include a security roadmap with prioritized recommendations.

During a Breach

All that advance planning and those security drills are put to the test when an incident actually occurs. Your organization's incident response team must spring into action when they identify a breach. They need to determine where the attackers are, what they seem to be after, how far they have advanced, and how long they have been in your systems.

As a CISO, you may want to contract security consultants to advise you on incident response. Such experts may be aware of newer and evolving security threats and have the knowledge, technology, and skill to defend against those attacks.

Just as important as responding to the attack is deciding how to disclose the incident to the outside world. Keeping the breach quiet may no longer be an option, given legal disclosure requirements and the likelihood that news of the event will become public. If you've been breached and you know it, you are in an absolute race against time to get your arms around what happened and what you are doing about it. However, you have to be careful about what you disclose, to whom, and when. Some companies opt to disclose as much as possible out of a sense of corporate responsibility.

Having a communications plan in place ensures that you release only relevant information to the correct authorities and the general public in a timely manner, using appropriate channels. It's important to work with your legal counsel as well as your public relations partners to determine not only how to put your best foot forward to customers, suppliers, employees, and the public, but also to determine what's required of you under relevant laws. Internal coordination with your legal counsel and public relations department is key.

While disclosure decisions are important, the larger concern during a breach is to gain control of the situation, remove attackers from your network, and get operations back to normal as soon as possible.

After a Breach

After operations are restored from a breach, one important question company executives should ask themselves is "Could this happen again?" They need to create a framework to identify gaps and vulnerabilities in their systems and decide what and how to improve. Changes could include hardening firewalls and upgrading security appliances that guard your email, endpoints, or mobile devices. If an intruder gained access to your network by first breaking into a partner's network, that vulnerability must be addressed.

Increasingly, companies are seeking improved intelligence about threats they face, since their core businesses do not encompass intelligence-gathering efforts. They often contract with a service provider that offers to identify various potential attackers and the methods they use to attack. To implement a cybersecurity framework, your company must adjust budgets as required and include input from C-level executives on how to meet the needs of each of their departments.

Machine learning is no silver bullet; they are only as good as the input they used to learn. If your machine learning algorithms are not well designed, the results won't be very useful. For example, false positives (prediction result saying it is true, but in reality it is false) prompting the CISO to trigger corrective measures unnecessarily. Similarly, false negatives (prediction result saying it is not true, but in reality it is true) will put the CISO in a very delicate situation where, even though the cyberattack is real, no action was taken. Unless the confidence level and prediction accuracy of the machine learning algorithms are improved, the trust factor to seriously consider machine learning as a counter-threat mechanism will remain questionable.

Chatbots are becoming prevalent in our professional as well as personal lives. The conversational interface is helping salespeople find specific data, helping customers directly resolve their queries without any intermediaries in between, helping us manage our diaries and schedules, and even helping security analysts craft proper responses to an incident. Yet, chatbots also bring risks. Microsoft introduced Tay, a chatbot to engage people through "casual and playful conversation," speak like millennials, learning from the people it interacted with on Twitter and the messaging apps Kik and GroupMe. Within 24 hours, a structured attack on Tay resulted in the bot shouting all sorts of misogynistic and racist comments.

The attack highlights another interesting yet highly vulnerable area. Just as businesses are beginning to adopt AI systems, attackers are also finding ways to manipulate these same AI systems. They are focusing on finding ways to turn AI against its owners, from hacking chatbots to deliberately misleading the pattern recognition algorithms.

These types of adversarial attacks means only one thing: the future of cyberattacks won't between humans, but between machines.

Questions for the CEO

As your business is increasingly become digital, are you at risk of losing control over the machines you are creating?

The answer should be both yes and no. The enterprise technology ecosystem is now much more complex, and much more open, than it used to be. Businesses can no longer control their own perimeters, because those perimeters no longer exist. At the same time, attackers have become much more creative and sophisticated. You can no longer rely on a set list of known attacks, because most attacks no longer fit recognizable patterns and signatures. On the other hand, that simply means you need to develop new definitions of control. Instead of attempting to control your perimeter against known attacks, you must learn to control unique attacks that have already breached your systems. There is still a place for prevention in cybersecurity. But you must extend your definitions of control to consider the full lifecycle of a threat to discover and mitigate threats before it is too late.

Does AI mean the end of the human element in cybersecurity or can they coexist?

They *must* coexist. There is currently a misconception about AI's role in cybersecurity. Many people hear the term AI and imagine a general AI system that can think and do things as creatively as a human. This general AI might be developed in the future, but at the moment AI's role in cybersecurity is very specialized. AI is good at processing a massive volume of threat data and providing a summary of what it found. But we still need humans to tell the AI what to look for, to review the AI's outputs, and to ultimately decide how to act upon the AI's findings. The goal of any AI-driven cybersecurity service is simple—let AI do what it does best, and let humans do what they do best.

Can hackers enter the machine and alter the path of AI?

There is one sobering truth about cybersecurity: it doesn't matter what defenses you put in front of attackers, they will eventually find a way around them. There is no "silver bullet" technology that will solve the problem of cybersecurity once and for all. Cybersecurity will always be a back-and-forth game between attackers and defenders that will constantly evolve as technology evolves. No matter how powerful AI is, it will not solve this problem. It is the next necessary step we need to take to secure our systems, but it is possible that attackers may one day find a way to bypass, or subvert, AI—the same way attackers have bypassed, or subverted, every other security measure cybersecurity experts have developed. Although it's important to ask this question, it is a moot point today. Attackers are already using AI to power their attacks, and we must deploy AI-driven defenses to keep up. In the future, we could be looking at dynamic machine learning algorithms that morph once they sense an intrusion, all without affecting the original purpose for which it was designed.

In the short-term, AI will become a standard element of cybersecurity. As enterprises further adopt the cloud, mobile, and IoT, they will open more vulnerability points for cybercriminals to exploit. As they further digitize, they will create opportunities for increasingly complex multi-channel attacks. Enterprise security staff are already flooded with more daily threat data than any human-only team can handle. This data will only increase every year, and so will the need for AI to effectively process it.

Conclusion

Cybersecurity is going to become more dependent on AI. In particular, as the notion of everything connected continues to expand, we can expect security risks from all possible systems, which the current controls won't be able to handle. There will be obstacles for sure. For example, AI needs to deliver greater accuracy in detection and fewer false positives for it to earn the trust.

AI powered cybersecurity is no panacea! They are not necessarily a perfect match either. Some attacks are very difficult to catch. There are a wide range of attacks, spread across ranges of time, and across many different systems. Second, the attacks are constantly changing; therefore, the biggest challenge is training the AI for all these varied scenarios.

In this chapter, we discussed the current state of cybersecurity measures and how AI can play a significant role in enhancing the security mechanisms of future systems. In the next chapter, we discuss concepts related to the Internet of things (IoT) and how AI can be leveraged to make "things" intelligent.

References

1. https://www.infosecurity-magazine.com/next-gen-infosec/ai-future-cybersecurity/

2. https://www.csoonline.com/article/3250850/security/artificial-intelligence-and-cybersecurity-the-real-deal.html

3. https://towardsdatascience.com/ai-and-machine-learning-in-cyber-security-d6fbee480af0

4. http://www.itpro.co.uk/security/30102/how-to-use-machine-learning-and-ai-in-cyber-security

5. http://www.information-age.com/using-ai-intelligently-cyber-security-123470173/

6. https://www.darkreading.com/threat-intelligence/ai-in-cybersecurity-where-we-stand-and-where-we-need-to-go/a/d-id/1330787?

7. https://blog.capterra.com/artificial-intelligence-in-cybersecurity/?

8. https://blog.capterra.com/artificial-intelligence-in-cybersecurity/

Intelligence of Things = IoT + Cloud + AI

With the Internet of Things (IoT) and artificial intelligence, one is the enabler and the other is the disruptor. Both of these technologies, individually, are transforming businesses and lives.

The Internet of Things (IoT) is a network of physical objects embedded with sensors and connectivity, so that they can exchange information with other connected devices in the network. In simple terms, IoT is about taking the inert and making it "smart," or even smarter, by connecting it to well, everything. It's exactly what it sounds like—things that have Internet! You take any object, embed sensors in it, attach a unique digital tracker to it, and then enable it to send and receive information without human interactions. You got your own IoT.

When objects can sense and communicate, it changes how and where decisions are made. Cheap sensors, improved wireless connectivity, and scalability through cloud computing have all made it possible to cost-effectively collect and process lots of data, analyze it, and act on it instantaneously. As a result, everything around us (animated, non-animated) is becoming connected

S. Mohanty and S. Vyas, *How to Compete in the Age of Artificial Intelligence*,
https://doi.org/10.1007/978-1-4842-3808-0_7

and smart. If you extend the concept of IoT to industrial scenarios such as manufacturing and supply chain management, you get the Industrial Internet of Things (IIoT). If you extend the concept of IoT to individual consumption scenarios in our daily life such as wearables, home appliances, personal assistants, smart thermostats, etc., you get Consumer IoT.

It is projected that IoT (and IIoT) will lead to the next-generation of smart operations at scale. For example, take a highly mechanical and physical intensive set up like a factory. It can be transformed into a "smart factory" where machines ("intelligent assets" or "smart machines") are capable of "talking" to each other and "making decisions" independent of human interference based on the data they collect and communicate. The result? Improved production times, improved cost-efficiency, reduced waste, and top-notch quality.

IoT doesn't only apply to M2M (machine-to-machine) communications; it extends to any task where close integration between people, processes, and machines (virtual and real worlds) is required.

With IoT alone, we can talk and listen to machines, but to understand machines and make them intelligent, we will need AI. AI provides us with the capabilities to learn and infer patterns from the vast amount of data collected through IoT.

What happens when we combine IoT and AI? It's a step-change transformation in everything—from machines to factories to automobiles to wearables to kitchens to buildings, and many more.

Let's explore a few scenarios to appreciate the transformative power of IoT and AI, combined.

Making Homes and Buildings Smart

A smart home usually consists of a central control unit (maybe physical or on the cloud) that reads information from every type of sensor spread throughout the home. Motion sensors, temperature sensors, and the like generate information every second and send it to the central unit. The central unit makes decisions based on the information sent to it by these sensors. Therefore, the entire home is connected and smart.

- Proximity sensors placed in refrigerators sense the stock inside and send data to the central unit. The central unit processes the data and places replenishment orders online based on your consumption patterns.

- Motion sensors in the house include facial recognition to recognize thieves and respond accordingly by sending a message to nearby authorities.

- Temperature sensors maintain the temperature of the home, according to the weather outside.

- Power consumption sensors keep tabs on usage trends and automatically turn devices/lights on or off based on the presence of people in the rooms.

Keeping Your Car Running

Imagine your car was fitted with all kinds of sensors, streaming data about the wear and tear, engine performance, and your own driving patterns, and that everything is collected through IoT gateways and pushed to the cloud. The AI running on the cloud continuously analyzes this data and generates predictions. AI reaches out to your Alexa when it thinks the car needs maintenance, and Alexa checks your calendar and talks to your favorite mechanic to find the earliest suitable schedule, and then asks you to confirm verbally. You'd never get into situations where you are oblivious to the fact that your car badly needs attention.

Keeping the Power On

When it comes to electricity supply, we are more affected by brownouts than blackouts. A blackout refers to complete power loss that impacts a large area, whereas a brownout happens when a sudden surge in demand forces your utility company to reduce line voltage in a few areas to manage the demand and supply better. The unintended consequences of a brownout are voltage fluctuations, which can severely damage electrical appliances.

Now consider home thermostats that sense the environment and temperature patterns and transmit data to your utility company, in real time. If it is an exceptionally hot day and there is a possibility of brownouts, the AI system on the cloud can see how many devices are operational, and then proactively turn the thermostat up a few degrees, while keeping the thermostat stable in temperature-sensitive facilities such as hospitals, thereby avoiding the brownout situation.

Making Sure You Stay Alive

There is more to the Apple Watches than a status symbol. Your wearables can do more than just read your heart rate and determine how much you have walked today. Imagine that your wearable sends an alert to the AI system when it detects that you might be having a heart attack. The AI system receives the alert, then goes into a mission-critical mode by triggering several actions in real time. It sends your location data to the closest ambulance, notifies the closest hospital to prepare ahead of time, alerts your doctor and recommends the fastest and shortest route to reach you or the hospital. Adding AI to IoT could mean the critical minutes that keeps you alive.

Making Sure You Stay Safe

We are concerned about the safety of our loved ones. This is the primary reason why we have gone to lengths to ensure our homes are protected. We install CCTV cameras all around the perimeter and intrusion detection systems that are connected to the law enforcement agency control room. Imagine if the data from your CCTV camera and your intrusion detection control system was continuously being pushed to the cloud, where the AI system was not only assessing the situation at real time but also applying vision recognition algorithms to correctly identify the intruder by checking against the criminals database and guiding the law enforcement officers on the next best actions.

Of course, there are data privacy concerns related to all of these examples. However, some people think the benefits clearly outweigh the risks.

The impact of IoT, the cloud, and AI, combined, requires a vivid imagination. We can apply the transformative power of these technologies to pretty much anything and everything.

IoT, the Hybrid Cloud, and AI Work Together

The examples discussed previously are no doubt are fascinating. But the key question is how would you build these kind of integrated applications? What are the technology building blocks?

The convergence of IoT and AI is simply not possible without an enabling platform and architecture. That's where the hybrid cloud steps in. All companies exhibit a unique technology landscape that has been built for and suited to their individual business needs and growth. The hybrid cloud is a disruptive technology and a business opportunity at the same time. To understand this disruption requires a deeper understanding of the converging architecture involving the hybrid cloud, IoT, and AI.

There are devices all around us that are collecting and transmitting data all the time. All of this data must be quickly analyzed for the next best action. When there's a delay, the data loses its value. This zero latency requirement necessitates incredible distributed collection and storage capabilities closest to the source. This means the edge of IoT must exhibit instantaneous capabilities to parse and analyze streaming data for the next best actions, at the same time pushing the bulk of the data to the cloud for deeper analysis.

AI, on the other hand, calls for immense compute power to operate on massive data sets. Speed and performance are additional considerations for AI systems, since decisions made by AI need to be fed back quickly to the IoT devices to make the predictions actionable.

The following list explains a few examples that highlight the AI specific requirements:

- Self-driving emergency response vehicles immediately respond to life-saving search and rescue operations such as floods, fire, etc.

- Medical devices can automatically defibrillate and send an alert notification to the nearest hospital.

- Financial crime detection systems include credit card swipes.

- On-demand recommendations for streaming video services.

- Apple's Siri and Amazon's Echo give instantaneous responses at the edge.

There are many more examples spread across industrial needs as well as consumer needs. In all these examples, one thing is common, the AI requirements not only involve lots of data but also real-time decisions (see Figure 7-1).

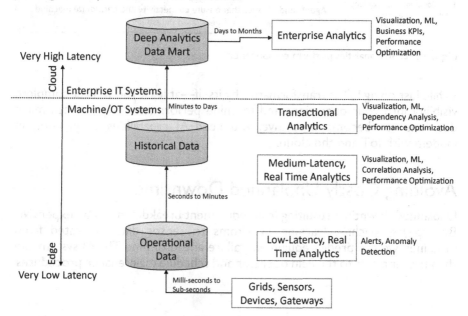

Figure 7-1. Data value chain and decision latency for IoT

Data collected by IoT can improve the prediction capabilities of AI; however, there are many tasks in the value chain (from raw data from devices to predictions, to outcomes) that need to be orchestrated well to deliver the final outcome. Figure 7-2 shows a conceptual view of the different components that make up an IoT analytics platform.

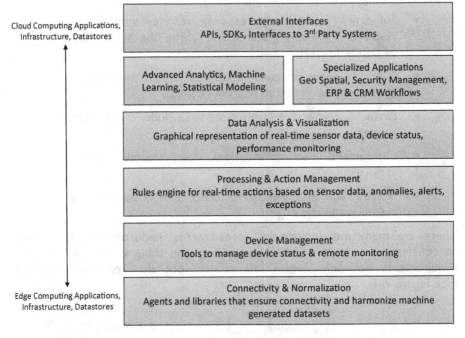

Figure 7-2. IoT analytics platform components

While just doing IoT is transformative by itself—at least you will get complete visibility into your operations and machine performance and you can make decisions—to become predictive, you need AI capabilities to function in tandem with IoT and the cloud.

Avoiding Costly Unplanned Downtime

Unplanned downtime resulting from equipment breakdown can be expensive. By applying machine learning algorithms to sensor data generated from machines, we can predict equipment failure ahead of time. The AI system can then trigger alerts to the field operator and schedule maintenance procedures.

Increasing Operational Efficiency

In factories, assembly line operations demand precision and strict quality controls adherence in everything. Take the example of spillage during the production process. Every 1% reduction in spillage can mean thousands of dollars in savings. By using IoT and machine learning, companies can significantly reduce leakages. Streaming data captured from sensors embedded in the machines can be analyzed at the edge itself in real time, thus sending alerts to human specialists to control spillage.

Enabling New and Improved Products and Services

IoT coupled with AI can form the foundation for improved product designs and in some cases even conceptualize entirely new products and services. For instance, by analyzing machine performance data, we can help spot patterns and gain operational insights—how is the machine working, is it able to perform based on the design specifications, is the real-world usage different from the usage considered during design, is there an opportunity to launch a new line of products, and so on.

Enhancing Risk Management

A number of applications that pair IoT with AI are helping organizations better understand risks and prepare them for rapid response, better manage worker safety, and cyber threats.

For instance, by using machine learning on wearable devices, we can monitor workers' safety and prevent accidents. Banks have begun evaluating AI-enabled real-time identification of suspicious activities from connected surveillance cameras at ATMs. Vehicle insurers have started using machine learning on telematics data from connected cars to accurately price usage-based insurance premiums and thus better manage underwriting risk.

The Role of AI in Industry 4.0

The prevalent notion around Industry 4.0 is to connect every asset and create digital environments. This is not entirely a true state. Industry 4.0 is more of an evolution than a revolution. There are several phases that you need to go through to achieve the desired level of maturity. The first phase begins with connecting all the assets together, which includes instrumentation, connectivity, and data collection. The next phase is to figure out how to make sense out of all this data. The final phase is to bring in automation to the entire lifecycle. Manufacturing agility is key to success for Industry 4.0 and AI has a critical role to play throughout the whole "create, make, and deliver" value chain to optimize and improve every process in order to achieve the best outcome regardless of external circumstances.

Industry 4.0 talks invariably steers into several technologies: edge computing, cloud computing, AI, IoT, platforms, etc. Even though each of these technologies has a role to play, the most critical components are its data and prediction capabilities.

Digital Twins

Modern manufacturing companies are leveraging IoT platforms to capture data from their machines and other sources to create a "digital twin," which is a digital model of an actual, physical object (see Figure 7-3). Digital twins enables us to visualize the real-life operational aspects of a machine. Sensors on the machine transmit data to its digital counterpart, mimicking the physical machine in action, then machine learning algorithms analyze the digitally transmitted data to optimize product performance and recommend the next best actions for the physical system. With digital twins, we have the ability to predict the physical world scenarios, optimize the asset performance, reduce maintenance costs, reduce downtime SLAs, and explore opportunities to monetize data and build new service offerings.

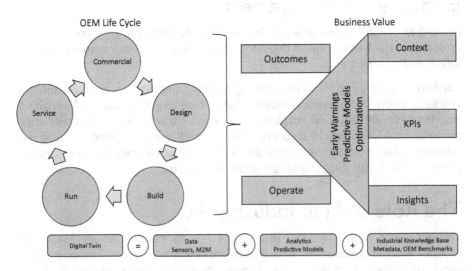

Figure 7-3. Conceptual view of digital twins

In essence, a digital twin is the convergence of the physical system, the digital mirror, and the underlying data that bridges them both.

How could a digital twin deliver transformational business outcomes?

First, digital twins—or rather the converging technologies of IoT, the cloud and AI—offer the possibility to optimize processes and apply effective usage of machines and resources. While these solutions are primarily aimed at lowering the cost of production, customers expect the end product to be inclusive of value-added services, like a dashboard, usage patterns, alerts for maintenance, etc.

Second, smart devices acquire extensive data about the real usage of the product at the customer location. When you analyze this invaluable data across your install base, you will be able to infer not only individual customer-install base patterns but also a world view of how other customers are using it, how successful they are, what challenges they are facing, and whether there are whitespace opportunities to launch new products or value-added services at a premium.

Third, because now you are not just selling a machine but a "machine as a service," you can change the entire business model in which your customers won't mind a premium insurance fee if you can guarantee the quality of their products and services. With digital twins, you can measure the health and performance of your machines. You are able to remotely manage the usage of these machines and predict possible outages. You are able to monitor machine utilization and recommend better capacity planning and production schedules, etc. All these insights mean that you can actually play a larger role in customer's business outcomes, and if you can commit to reduced downtime SLAs, surely your customers will be willing to pay you a premium.

How would you create a profitable digital twin strategy for your business?

If you are purely an industrial manufacturing company, you have three options:

- *Become an enabler:* Start developing and embedding IoT technology, such as endpoint networks, edge computing, and cloud infrastructure, into your products and offerings. This will enable you to cross-sell pieces of digital twin components to your customers and gain market share. Your customers then have to take external help of building in-house capabilities to architect digital twins relevant to them. However, on the cautioning side, beware that this market will be dominated by a small number of global giants who can deliver the entire stack at a fraction of the cost.

- *Become an engager:* Besides manufacturing world-class products equipped with sensors and compute power, you can venture into designing, creating, integrating, and delivering value-add services to your customers. The value-add services could be like adding a real-time dashboard, trends of historical machine performance, alert notification based on thresholds, etc. These offerings may not give you a distinct advantage because traditional software companies specialize on delivering such capabilities; however, you will certainly be able to develop customer intimacy by offering digital services over physical machines.

- *Become an enhancer*: Here, the goal is to provide enriched end-user engagement and offer new services using the data from the customers themselves and third-party sources. This is where digital twins, monetizing data, and new revenue models come into play. The value-added offerings need to be machine specific and industry specific to raise the entry barrier for others to replicate.

What Is Your Intelligence of Things Strategy?

The use cases involving AI and IoT convergence are evolving so rapidly that in a short span of time, almost all "dumb" devices will have to become intelligent.

Here are a few considerations to keep in mind for your intelligence of things architecture:

- There will be an ensemble of many IOT devices and they need to communicate and work together throughout the organization. Depending on the product and services, this ensemble may be with a person (with wearables or personal devices), a house, a vehicle, a project, or a factory.

- The number of devices and the number of manufacturers will proliferate. Consequently, there will be no standard way to achieve machine-to-machine communication. Hence, choosing the right cloud partner will be crucial so that the devices can work together in a roundabout way of machine-to-cloud-to-machine. There will be a need to establish "APIfication" of devices as well.

- An integrated view of data across IoT and other enterprise data sources, as well as external data such as weather patterns, is required to achieve the full potential of prediction and autonomous decision.

Similarly, here are a few considerations to keep in mind for identifying the opportunities for innovation with IoT, the cloud, and AI:

- *Customer experience*: Identify possible innovation on the customer experience side by reimagining how a customer uses a product or service and how the ultimate value to the customer is created, distributed, consumed, and serviced. Is there an opportunity for a better customer journey by collecting new data, adding to the mix of existing processes data and creating new processes and partnerships?

- *Product and services*: Identify possible opportunities for a product to provide more value for users by leveraging a combination of IoT, the cloud, and AI (with the mix of infrastructure, processes, policies and people). Imagine new features to the product or enhance current features to perform better. This is more about the function and form factor, and not about the enabling technology.

  On the services side, identify the opportunities on the service and the ultimate consumption by customers, i.e., how, what, and when a service is offered and consumed. Imagine if a product were to be transformed into a service.

- *Enabling technology*: Innovation on the enabling technology is not necessarily specific to a product or service, but it's more about the adoption of technology to deliver on the reimagined customer experience, product, or service and business model. This is where the mix of IoT, AI, and the cloud and its adoption for your specific business needs to be carefully evaluated.

- *Business model*: Identify the opportunity to enhance the current business model to create and deliver value to existing customers (or entirely new customer base), by putting it all together. Imagine a new customer experience with new or enhanced products and services created by adopting and enabling a technology mix of IoT, the cloud, and AI.

Figure 7-4 shows AI and IoT convergence use cases across industries.

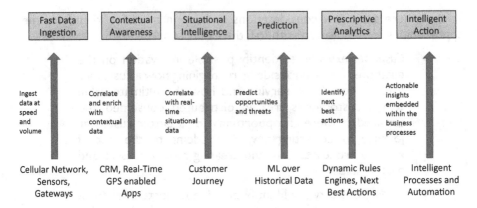

Figure 7-4. IoT analytics value chain

Here are a few more industry-specific examples.

- *Aircraft*: For airlines, zero downtime means higher revenues. Today, aircraft engine manufacturers are fitting millions of sensors, and the main goal is to not only understand how the aircraft is performing during each flight but also to become predictive about engine wear and tear. The result? Higher safety and fewer downtimes.

- *Oil rigs*: The machines deployed for drilling are capital intensive, thus the oil companies must keep improving on their operating costs. When these machines break down, companies incur huge losses, but it's not economically viable to keep expensive machines on standby. The solution? Make the machines smart so that their utilization, performance, and conditions can be continuously monitored and analyzed. Predictive maintenance and conditions based monitoring significantly reduces operating costs.

- *Manufacturing*: Manufacturers are investing to make their factories and plants smart so that their machinery and assembly lines can lead to the creation of autonomous factories in the future.

While at one end, the convergence of IoT, AI, and the cloud enables businesses with stronger value propositions, both in the B2B and B2B2C space, there are several challenges that businesses need to be aware of, especially in the context of managing the volume, velocity, and variety of data. Every disruption establishes newer opportunities and markets that did not previously exist; however, instead of taking the plunge, you need to assess what is relevant to your industry and define your own intelligent of things strategy.

Best Practices to Develop Your Intelligence of Things Strategy

Develop and Articulate Your Own Value Proposition

IoT and AI, combined, present a broad range of transformative opportunities. However, you need to carefully evaluate use cases before succumbing to the lure of putting sensors in everything and expecting magic to happen. IoT as a technology is still maturing; there are issues in connectivity all the time, and sensors may get damaged or malfunction due to harsh weather conditions. They may also just stop responding due to power fluctuations, hence success of your IoT initiatives depends on how well you have evaluated the whitespaces, complementary technologies available in the market, and the problems you are going after.

You need to develop a good understanding of the megatrends and competitive forces at play in your industry. What are the analysts saying about the application of IoT in your industry? At which point in the value chain are the customers getting frustrated? What additional data or which events, if made responsive, will significantly improve customer experience? Taking these data points, you need to develop your own SWOT analysis and business cases for your IoT initiatives.

Evaluate Customer Needs

For any business, it is crucial to get as many details about customer needs (explicit or implicit) as you can. In case of IoT, being an unchartered territory and novel, you need to go beyond the typical customer survey type of approaches and adopt a wide range of techniques, such as customer personas and customer journeys, to develop a roadmap of offerings and services that will be well accepted in the market.

Conduct Value-Chain Analysis and Profitability Analysis

The next step is to create a value-chain analysis and profitability analysis of your industry. Instead of taking a narrow constraints based view of your business, you should take a broad view of the industry. In some cases, this may result in diversifying to create a completely new line of business, but that is the key to come up with a competitive edge and differentiations.

Collaborate to Partner

It would be foolish to think and act as if you will have the wherewithal to do everything on your own. The technology is evolving at a rapid pace, customer needs are changing every day, and the market trends have significant implications. It is important to understand the solutions providers in your market and monitor their progress and challenges. This map will provide you with enough clarity to collaborate and partner in scenarios where you can gain speed to market with less cost.

Evaluate the Technology and Do a Fit-Gap Analysis

There are many vendors and solutions in the market. Some are very narrow but do the job very well, some are broad and offer greater flexibility to try new things. It is important to understand all these offerings and develop a fit-gap analysis specific to your goals and objectives. If your organization is new to the field of connected devices, your success largely depends on how quickly you can launch pilots and establish ROIs for business functions.

IoT, AI, and the cloud are fast-evolving technologies, hence your technology architecture needs to be flexible and componentized. In the overall architecture, the components may seem to be low in maturity today, but tomorrow they will attain sufficient maturity or some components may completely become outdated. Hence, it is important to keep an eye on the technology trends, including how fast they are evolving, what other complementary technologies are coming into play, and how cost-effective they are.

We have listed a few important areas you need to focus to develop a robust technology roadmap covering data, analytics, recommendations, performance, and overall cost-effectiveness.

For example, under "Insights," it's important to answer questions like the following:

- What data would give you clear views about your product usage and performance tracking?

- What data would be valuable for your business functions?

- What data would enrich the customer experience?

- What additional data do you need to collect to deliver these insights?

Analytics questions may include the following:

- What insights, if embedded into your products/offerings/processes, will make your company more responsive to customers or market scenarios?

- How complex will the "math" be? Do you have to buy specialized packages like optimization libraries or deep learning libraries?

- How would you manage and administer these insights?

Performance questions might include these:

- What is the data processing performance criteria?

- What are the consequences of not doing data processing at the edge (versus moving everything to the cloud)? If the cloud is the answer, would you lose a few key actions?

- How real time would you like your offering to be?

Operating requirements questions might include these:

- What operating conditions (temperature, moisture, pressure, access, and vibration) will you focus on?

- What are the different scenarios you will enable your security and access control for? The rest you will leave to customer-specific asks.

As you build your technology roadmap, you need to put on a pragmatic lens. Simply creating an intelligent solution will not bring you success, especially if the cost of building such a solution outweighs the commercial and implementational easiness aspects.

Build Your Intelligence of Things Roadmap

Once you've done all of the mentioned activities, you will have all the desired components to develop your intelligence of things roadmap (see Figure 7-5). The roadmap will help you plan and communicate to the stakeholders within your company, your partners, and employees the timelines, initiatives, pilots, changes, and expected outcomes.

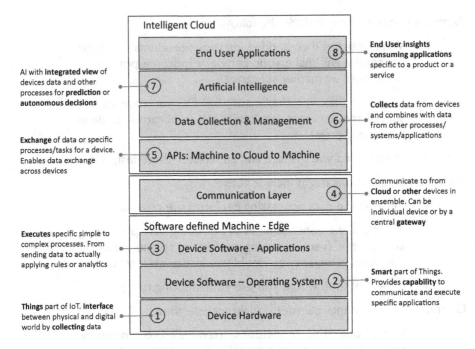

Figure 7-5. Multi-layered intelligence of things architecture

One of the best ways to develop your roadmap is to embrace Amazon's flywheel strategy. Start with a big vision, not necessarily a big capital intensive and complex bet. You need to start small, focusing on pilots that are easy to experiment to test your thinking. This can further evolve to create a minimally viable product to launch early in the market and gain the early movers advantage or you can identify customers who are willing to co-innovate with you in a profit-sharing model.

There are three methods that can help you articulate your roadmap to your business stakeholders, employees, and customers:

- *The future press release:* Start with the end in mind and develop a press release for your product or service offering. Since this is going to be a market facing announcement, you will be forced to articulate the uniqueness of your offering, which will in turn help you to solidify your vision.

- *An FAQ for your plan:* Come up with the potential questions you may face from the market, investors, employees, business stakeholders, and partners. The frequently asked questions (FAQ) and corresponding answers will help you socialize your offering in a much more acceptable way than a black box that people need to figure out by themselves.

- *A user manual:* Successful companies believe in a crowdsourcing model, wherein the users can create more and more utilities using your offering. This helps in significantly gaining market share in less time. Develop a user manual, DIY videos, APIs, and tutorials for end users to build smart apps.

Conclusion

What does the advancement of new technologies like AI, IoT, and the cloud mean for the vast number of people employed in industrial manufacturing setups?

AI, IoT, the cloud, and the other emerging fourth industrial revolution technologies are here to stay. They are changing forever the way things are designed, manufactured, and delivered. These technologies are making everything smart in the entire manufacturing lifecycle (buy, make, sell, and service) with additional complexities of managing smart devices, smart agents, and in some cases completely autonomous processes.

The key question is, will the role of managers, technicians, machine operators, and factory floor workers change dramatically? If yes, companies need to find ways to monitor and track activities in which more and more human-machine collaborations will become the norm to executing projects.

In this chapter, we discussed the art of possible by combining the capabilities of IoT and AI to improve our way of life as well as transform industrial sectors. In the next chapter, we discuss how AI can improve and transform IT operations.

References

1. https://www.wired.com/insights/2014/11/iot-wont-work-without-artificial-intelligence/

2. http://www.information-age.com/innovation-mining-iot-ai-monitoring-technology-123470678/

3. https://www.fungglobalretailtech.com/research/deep-dive-future-customer-experience-ai-iot-retail/

4. https://www2.deloitte.com/insights/us/en/focus/signals-for-strategists/intelligent-iot-internet-of-things-artificial-intelligence.html

5. https://www.strategy-business.com/article/00294?gko=a9303

IT Operations and AI

Every business is becoming a digital business, and as a result, there is a continuous drive to put everything on the cloud, become data driven, embed AI into every process, and do everything in an extreme Agile model. While all these are good and probably an essential complexity of becoming a digital business, at the same time being digital means being caught between the constant tussle of development and operations activities. This is reminiscent of the second law of thermodynamics: entropy increases over time unless energy is introduced into the system from outside.

Let's explore this analogy further and see how AI can play a crucial role in the world of IT operations.

The IT operations world—and especially the production run, maintenance, and care activities—are in a constant fight against entropy, which refers to disorder. IT operations personnel spend a significant amount of time and energy trying to keep everything up and running and are always under mounting pressure to predict when things will go wrong and be prepared with fixes and lengthy root cause analysis documents.

In contrast, the development activities actually *increase* the level of entropy. Every new app moved into production, every new release of underlying OS and associated upgrades, every new technology introduced to the users, is a change that has to be made, increasing the probability of disorder that the operations team fights so hard to prevent.

© Soumendra Mohanty, Sachin Vyas 2018
S. Mohanty and S. Vyas, *How to Compete in the Age of Artificial Intelligence*,
https://doi.org/10.1007/978-1-4842-3808-0_8

While the fast pace of development world is inevitable, if the change is unmanaged, it will eventually cause something to break, unless people put in continuous effort to prevent those breaks. These conflicting goals, taken to extremes, can snowball into an unhealthy situation within the enterprise, with operations actively fighting against the change and development.

What is needed? A flexible, holistic framework consisting of a set of diagnostic and predictive tools, automation, and humans-in-the-loop capabilities that will enable the operations teams to embrace change. This flexible holistic framework is called *AIOps,* a term coined by Gartner Research to describe a new way of operating IT based on augmenting human skills with automated analysis enabled by AI techniques.

The term AIOps itself is neither a solution nor a philosophy, but an overarching framework for how to address increasing performance demands and manage IT complexity. This involves the development of new and powerful technologies that have the potential to transform IT operations and digital businesses at large.

Gartner proposes a plethora of capabilities that need to be integrated and managed:

- Historical data management, inclusive of structured data and unstructured data

- Streaming data management, inclusive of multi-device access and usage

- Log data management primarily generated from systems and networks

- Automated pattern discovery and prediction

- Root cause analysis and scripts serving as hot-fixes

- Software-as-a-service, primarily cloud-based solutions and associated data services consumed in a one-to-many model

Today's IT landscape is too complex for any single person or even a team to manage on their own; instead, the skills required consist of multiple specializations (full-stack engineers). In turn, each of these specializations requires its own specific tools to monitor, guide, recommend, fix, and enable IT operations personnel to shoulder the responsibility effectively. At a certain point, it becomes very difficult and chaotic to combine these partial views to gain a holistic understanding of what is actually happening in the environment.

Because of the scale and nuances of modern IT environments (hybrid cloud environments and everything-as-a-service type of landscapes) and the increasing velocity of change that businesses require, old-style rules-based and the keep-bubbling-up approach (L1-L2-L3) have become unsustainable and brittle, requiring more effort and care. Especially as infrastructure becomes more self-modifying (infrastructure-as-a-code) in response to changing conditions and demand, IT operations needs to accelerate just to keep up.

With every new technology, there is a need to understand the "why" behind the what for better decision making.

Why Does Your Business Need AIOps?

Digital transformation hinges around four broad areas and organizations may have different implementation strategies across these areas:

- Hardware to applications to business transactions (impacting revenue, cost, or customer satisfaction)

- Legacy technology to new-age microservices transformation (impacting dual speed of IT and responsiveness of IT systems)

- Infrastructure consolidation and modernization for on-field (things), on-premise, and on-cloud (impacting robustness, scalability, and performance)

- Different time intervals for changes across all of the above (impacting go-to-market readiness)

It is clear that digital transformation requires increased cloud adoption and readiness to respond to rapid development and production deployment capabilities. However, when you add essential complexities like demand for continuous innovation and continuous development (CICD), increasing adoption of machine agents, Internet of Things (IOT) devices, Application Program Interfaces (APIs), and so on, your IT operations scope is exposed to too many unknowns. This strains the traditional service management best practices, the team's capabilities, and the skills and tools to the breaking point. See Figure 8-1.

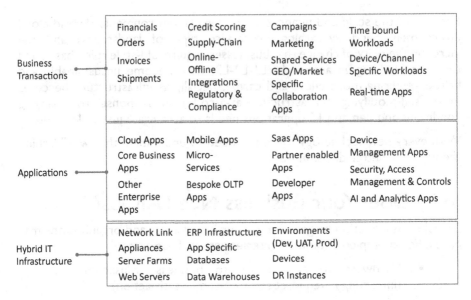

Business Transactions	Financials Orders Invoices Shipments	Credit Scoring Supply-Chain Online-Offline Integrations Regulatory & Compliance	Campaigns Marketing Shared Services GEO/Market Specific Collaboration Apps	Time bound Workloads Device/Channel Specific Workloads Real-time Apps
Applications	Cloud Apps Core Business Apps Other Enterprise Apps	Mobile Apps Micro-Services Bespoke OLTP Apps	Saas Apps Partner enabled Apps Developer Apps	Device Management Apps Security, Access Management & Controls AI and Analytics Apps
Hybrid IT Infrastructure	Network Link Appliances Server Farms Web Servers	ERP Infrastructure App Specific Databases Data Warehouses	Environments (Dev, UAT, Prod) Devices DR Instances	

Figure 8-1. IT systems topology

Here we have listed a few key friction points that exist in the current service management set up:

- *Very little automated approach to IT operations*: Modern IT systems landscape include managed cloud, hybrid cloud, third-party services, SaaS applications, mobile applications, APIs, crowd-sourced applications, IoT devices, etc. In short, the IT environment is complex, dynamic, and elastic in nature. The traditional approach of managing IT environments through human oversight is no longer a feasible solution. With too many touchpoints and multi-channel access to systems, applications, and data, the need is to have a highly automated way of tracking, monitoring, and resolving issues.

- *Too many systems, too much data to deal with*: The IT systems landscape is no longer confined to large monolithic stacks; there are way too many specialized applications and systems solving specific enterprise needs. The result? Exponentially larger numbers of events and alerts and increased service ticket volumes. It has simply become too complex to understand all the data generated from all these applications and serve the SLAs at the same time. With too many touchpoints and multi-channel access to systems, applications, and data, the need is to have a highly automated way of tracking, monitoring, and resolving issues.

- *Need for almost real-time responses to events*: Going digital has created a divide in the enterprise IT systems. Frontend customer facing systems are running at a higher speed compared to backend process-centric mission-critical systems. In many organizations, the IT's role in enabling business has been so phenomenal that now IT has become the business. Subsequently, "consumerization" of technology has changed user expectations to such an extent that businesses now have to respond to IT events much faster than ever before. Whether it is responding to real or perceived situations, the responses need to happen almost instantaneously.

- *Compute infrastructure moving to the cloud or to the edge*: Cloudification has simply disrupted the very existence of IT functions, budgets, and controls. More and more compute infrastructure needs are now on a usage-based principle. While this trend is good from a cost management perspective, it creates additional strains on IT teams to manage hybrid environments, where some applications are on cloud and some are on-premise.

- *Agile development methods at odds with Operations responsibility*: DevOps and Agile methodologies have enabled the businesses to experiment, prototype, and deploy applications to production at a rapid pace. However, the responsibilities of managing, monitoring, and maintaining these applications are still left to the IT operations teams. Lack of knowledge about these new technologies is pushing the IT operations teams to be in a catch-up mode all the time.

- *Too many false positive alerts*: In the digital world, owing to the nature of business and numerous customer touchpoints, the business applications and IT systems are at the receiving end of increased user activities all the time. There is virtually no down-time expected. This heightened usage requirement consequently is subjected to an increased number of false alarms in terms of response times, accessibility, availability, and scalability. IT operations teams are challenged to analyze an increasing volume of individual tickets and pinpoint to the root causes. They have to deal with multiple monitoring solutions and multiple related alerts for the same issue that could also create an event storm that seizes up resources when only one notification and incident ticket would be more practical.

Moreover, once the root cause is identified, the IT operations team goes through a difficult triage procedure to route the issue to the right team based on earlier guidelines about who will fix the issue; this particular aspect is cumbersome because the IT landscape no longer consists of a few large monolithic applications. Rather today's IT landscape is all about a diverse set of infrastructure (on-premise and cloud) apps, varying SLA expectations, and multiple vendors (some of them open source).

So, how can AIOps help solve these kinds of issues in an intelligent, integrated, and automated way?

What Constitutes an AIOps Platform

IT operations personnel are overwhelmed with tickets, stringent SLAs, and numerous data points they need to analyze in order arrive at the root cause. They see that traditional IT operations management techniques are not up to the task, and they don't know where to start. They expect AIOps systems to figure it all out. Two key expectations from AIOps are automation and prediction.

Automation

Enable the IT operations professionals to automate what they are currently doing manually. Through automation they can speed up issue identification and resolution activities, and therefore also solve more issues in a given time or with a given set of resources.

Algorithms can automate the process of analyzing and correlating event data, which if done manually, would take hours. Millions of events can be reduced to tens of incidents automatically, using sophisticated algorithms that can de-duplicate, filter out the irrelevant data, and correlate event feeds in real time.

To further elaborate on the automation aspects, let's consider a few examples: correlating customer profile information with financial processing applications and transactional data to identify outliers and highlight performance impacting factors; evaluate unstructured data in service tickets to identify problem automation candidates; categorize workloads for optimal infrastructure placement; and correlate incidents with changes, work logs, and app dev activities to measure production impact of infrastructure and application changes.

What do all these tasks have in common? To efficiently perform these tasks, the key requirements are reasoning capability and contextual knowledge. The IT operations personnel need to correlate among several data sources to identify outliers, identify causal relationships, and refer to system logs to surface the root cause. Quite simply, the expectation is for a system to take over these manual investigation tasks to highlight as many possibilities as possible, then the SMEs can apply the domain knowledge necessary to reason and arrive at the next best steps.

This level of automation means that incidents can be detected instantly without requiring humans to manually connect the dots across various tools and application silos. For example, the AIOps platform can curate every incident observed in the past and capture all the tribal knowledge inclusive of scripts executed to resolve that incident. Should a similar incident occur in the future, algorithms can be used to automatically refer the knowledge base and reuse the approach taken to the incident. The role of AIOps is to increase the productivity, responsiveness, and productivity of the IT operations personnel by automating the manual tasks that they perform.

Prediction

By analyzing past incidents and the current stream of events, the system should be able to predict several things: a) determine the severity of the issue, b) categorize it into a kind of problem, such as infrastructure related, application related, user administration or access related, etc.

In other words, the expectation is that there should be some sort of intelligent system that continuously works on the data and, by learning what is normal behavior and what is an abnormal behavior, raises appropriate notifications, provides recommendations for fixes, and highlights confidence levels against those recommendations. If the intelligent system achieves significant confidence levels, then it may be smart to autonomously execute the fixes.

Keeping these two primary expectations in mind, let's discuss the capabilities required to develop the AIOps platform:

- *Ability to integrate extensive and diverse IT data sources*: The platform should be able to break the currently siloed tools and IT disciplines and integrate data related to events, metrics, logs, job data, tickets, monitoring, etc. across the entire stack of infrastructure, applications, and end-business transactions. The aim should be to create an end-to-end topology.

- *Ability to aggregate and provide a holistic view*: The platform should be able to aggregate historical data for trends analysis and learn and at the same time capture streaming real-time events.

- *Ability to compute and perform advanced analytics*: The platform should be able to compute and employ sophisticated machine learning algorithms, including NLP capabilities to eliminate noise, identify patterns, isolate probable causes, expose underlying problems, and achieve other investigative and prediction oriented tasks. The platform should also have the ability to simulate different scenarios in real time to enable the IT operations experts to carry out "what-if" analysis and speed up the triage process.

- *Full stack visibility/ability to visualize complex patterns in data*: The platform should have the ability to visualize various interdependencies and relationships in vast amounts of diverse data. This is required to help the IT operations personnel quickly analyze and understand the interdependencies between systems.

- *Ability to automate tasks*: The platform should be able to automate tasks, taking the outcomes generated by machine learning, to fix the issues.

Let's now explore how the AIOps system will be able to solve the common IT operations tasks.

Forecasting:

- *Task*: Determine when prior defined thresholds will hit the circuit breakers and perform "what if?" analysis to find root causes.

- *Data*: System logs data, application performance data, user concurrency and usage patterns, and events.

- *Algorithms*: Linear regression, change detection, seasonality decomposition, and Box-Jenkins.

- *Machine learning*: Mostly supervised; the system is already trained on historical data.

- *Visualization*: Graph analyzers to help quickly spot dependencies and linkages.

Example of actions: plan for in-time provisioning of resources in anticipation of blackouts.

Probable root cause analysis:

- *Task*: Correlate a diverse set of data points to automatically identify a small number of potential causes.

- *Data*: System logs data, application performance data, user concurrency and usage patterns, events, and knowledge bases consisting of root causes and corresponding fixes.

- *Algorithms*: Pearson correlation coefficients or other suitable linear correlation algorithms.

- *Machine learning*: Unsupervised learning approach leveraging historical data and knowledge base without any prior training or specific goals.

- *Visualization*: Recommendations with specific actions, corresponding confidence scores, decision trees, etc.

Example of actions: Redirect attention and resources to respond to the right cause.

Clustering:

- *Task*: Find similarities and frequency distributions of word pairings to categorize various issues according to severity and root causes.

- *Data*: System logs data, application performance data, user concurrency and usage patterns, events, and knowledge bases consisting of root causes and corresponding fixes.

- *Algorithms*: Levenshtein distance and Latent Dirichlet Allocation for topic modeling.

- *Machine learning*: Unsupervised learning.

- *Visualization*: Hierarchical clustering and self organizing maps.

Example of actions: Identify the similarity of a new issue with earlier ones and redirect to the right expert or respond automatically.

Continuous improvement:

- *Task*: Determine future behavior of operations metrics and system performance by continuously monitoring daily, weekly, and monthly behaviors.

- *Data*: System logs data, application performance data, user concurrency and usage patterns, events, and knowledge bases consisting of root causes and corresponding fixes and implications due to new technology adoption.

- *Algorithms*: Bayesian belief networks and stochastic non-deterministic algorithms.

- *Machine learning*: Mostly unsupervised approach and deep learning to continuously learn from data.

- *Visualization*: Recommendations with specific actions, corresponding confidence scores, decision trees, etc.

Examples of actions: plan for provisioning resources to avoid blackouts.

What's Your AIOps Strategy?

IT operations has always needed to ensure stability and availability in IT performance management. This has led to cautious adoption of leading edge technologies like machine learning (ML). Implementation has typically been by vendors in tools targeting specific use cases.

A comprehensive, enterprise-wide analytics and machine learning strategy has eluded most IT shops. The challenges in tooling, knowledge, and expertise have been too steep and there are too many other more pressing demands. In digital transformation, IT becomes the business. IT needs to balance the traditional demands of performance management with those of innovative business. To do this, IT must bridge IT and business data silos and leverage machine assistance to manage evolving customer desires and technological change.

Mobility, the "consumerization" of customer expectations, elastic cloud infrastructure, rapidly changing application technologies, and continuous delivery all place new demands on IT that cannot be met with the old approach. Big Data has matured and commoditized sufficiently to support the vast and growing quantities of real-time data and processing required. The general adoption of APIs and open data platforms enables the sharing of critical information that used to be siloed in tools and databases. Machine learning is mature and ubiquitous in our technology and culture.

Gartner estimates that by 2020, 50% of enterprises will actively use AIOps platforms to provide insight into both business execution and IT operations. IT doesn't necessarily need to change what they do; they just need to do it faster, with less manual intervention, and across a broader and more complicated ecosystem.

The following sections discuss the strategies we recommend.

Establish Tools and Processes to Self-Organize the Systems Topology

As the change happens across various areas of operations over time, it's important to deploy tools that can monitor and discover the entire IT systems topology and its relationship on a continuous basis and keep the integrated topology view up-to-date. Such a dependency view and knowledge is key to faster discovery, faster root cause analysis, and better assessment of hotspots/bottlenecks in the IT systems landscape.

Establish Real-Time Data Exploration and Reporting

Service desk ticketing tools typically provide managers and executives with reporting capabilities. Most implementations come with a set of out-of-the-box, boilerplate reports with the ability for administrators to build new, custom reports and some ability for individual report users to change views and parameters to suit their needs. Most now also offer scheduling functionality and the ability to generate sharable presentations out of reporting queries. This is sufficient if your organization has the discipline to have scheduled reviews, where everyone brings their reports to the table and you can spend the time correlating, corroborating, reviewing, discussing, and leaving with a measurable action plan. If the demands of your business, market, and competition don't permit this use of your time and resources, it has substantially less utility. A better approach is to have all necessary data available in an easy-to-use exploration tool that's available to everyone in the organization and updated in real time. This allows all stakeholders to see the top level (executive) view and dynamically investigate interesting trends, anomalies, or issues together, without having to collate reports or run new ones.

Natural Language Processing (NLP) to Eliminate Manual Categorization

Reporting and incident management rely on the fundamental organizational principle of categorization. Every ticket has structured properties assigned based on the source, process, etc. Routing, ownership, and ultimately reporting rely on user-, system-, or triage-assigned categories, which are often a reflection of what is in a free-text field like "Description" or "Issue."

In IT, it is a known but rarely discussed fact that this model of classification is broken. Users don't have the patience, discipline, or knowledge to accurately self-assign categorizations and priorities. Categories represent an inside-out view of the business: what IT thinks is important is to treat the issues, not what users think is important to say about them. IT departments build category hierarchies once and iterate on them over time without rationalizing or pruning options. Any sufficiently mature IT shop will have hundreds or thousands of possible categorizations, with entries that are too broad, too narrow, meaningless, or irrelevant. A top category will always be "Other." Additionally, static classifications, whether assigned by the user or the system, can't capture what is new. As a result, using categorization for reporting or intelligent routing is practically impossible. The remedy most often employed is simple brute-force reclassification: staff or contractor resources read ticket descriptions to correct or add categorizations. This is unnecessarily expensive, inconsistent, and inefficient.

A modern, analytics-based approach to address this problem would use natural language processing (NLP) backed by machine learning to process the free text of the tickets. Categories could be applied, but a more efficient approach would be to bypass categorization altogether and have dynamically updating and changing issues clustering based on what is actually happening in the environment. Natural language processing is all around us. There are simple applications like predictive typing on your mobile device and grammar checking in your word processing program, to complex implementations like auto-correct and de-duplication in Google search or résumé processing in Taleo. Chatbots, the next "everyday" implementation of artificial intelligence, rely on NLP to work. NLP needs to be a pillar of every IT ticket and incident management strategy.

Implement ML Across the Enterprise

Machine learning (ML) is a type of artificial intelligence (AI) that gives computers the ability to learn without being explicitly programmed. Said another way, ML is a capacity of computer systems to learn from data and computational results without direct human intervention. Understood this way, ML is not about doing or generating something completely new. It is about building on a foundation of domain knowledge and practices to automate adaptation in that context. In IT, we want to use ML to mimic what our best domain practitioners do, but do it faster and more consistently, simultaneously freeing up those practitioners for higher-value work. Take the NLP example from the previous strategy. If we use machine learning on NLP processing of free text fields, we don't expect the system to learn without a language dictionary and grammar, nor without a list of words that can be ignored (e.g., "or," "the," etc.). Programmed with this information, the system can then begin to learn from the data which words appear most frequently and in what combination, how variations of the same thing can be expressed, and even the context to determine when words are misspelled.

These are all things that humans can do. Implementation of ML in an IT context should be done with the goal of automating human tasks that involve learning. These are typically concrete and focused: monitoring activity and alerting/notifying, triage, routing, probable cause investigation, capacity modeling and prediction, etc. Implementing ML to take over responsibility for these tasks is manageable given the state of technology and IT domain knowledge. Doing so will free up your human capital for other value-added activities.

Bridge Data Silos and Disciplines to Unlock Hidden Insights

Previously, we talked about the virtues in moving from static reporting to dynamic data exploration. The greater amount and diversity of data that can be included, the more powerful and meaningful discovery and analysis becomes. It is important to not just bridge IT data silos (service management and operations), but also to bring together data from IT and business.

Fundamental to this approach is creating an integrated and collaborative relationship between incident management and IT operations. Operations has its own challenges in bridging the divide between performance management tools, but the need to connect the real user experience captured in service tickets with performance metrics and events can't be overlooked.

The two levers IT has to pull as ticket volumes and complexity increase are to a) get faster or b) prevent issues in the first place. Traditional strategies, analytics, and tools are typically focused on a. This is necessary, but still reactive. A few IT shops have an effective problem management practice. Lack of staffing, expertise, resources, or insights, exacerbated by organizational silos with different incentives, means robust problem management is rare. See Figure 8-2.

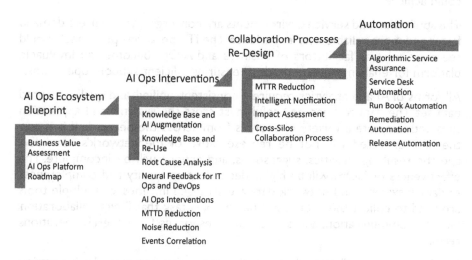

Figure 8-2. Illustrative roadmap for AIOps implementation approach

The strategies discussed will help create shared understanding and collaboration that will dramatically improve your organization's ability to respond. That, however, will not be enough to deal with the scale, speed, and unpredictability of digital transformation. IT must leverage these strategies to actively engage in proactive problem identification and remediation. Machine learning and cross-pillar analytics will reveal common problems driving incidents and create the space for organizations to engage in preventive automation and process improvement. This must be part of any incident management analytics initiative.

Conclusion

There are no quick, easy solutions to enable a business to match the pace of digital transformation, let alone be successful at it. Tasked to keep the lights on, IT operations has always been a conservative practice, because the primary business expectation is that the IT operations team should only support the business—nothing more, nothing less. However, if a business is to digitally transform, the IT team must transform as well.

The IT operations teams must learn more about their environment, the new technologies, and their implications. The learning mindset will encourage IT operations professionals to move away from a "support" only task to an Agile participant. The potential of automated and machine-assisted collection and interpretation of data are foundational capabilities of AIOps that can deliver results that far exceed what the IT operations team working on their own could achieve.

The application and service environments are converging as end users demand faster and more reliable user experience. The IT operations personnel should see their roles as facilitators of business and AIOps becomes an invaluable platform for enabling new methods of ensuring efficient support operations.

All forms of AI and analytics require a consistent, unified, and quality assured data set. The data to become useful must cross the systems and application silos and integrate a variety of sources from across the network. If it is not, the data collected will not be representative of the network's condition, and the resulting analytics, suggestions, and actions will be incomplete. The effectiveness of AIOps will be highly dependent on quality and completeness of data. If systems and network data are arranged in siloes, or multiple tools are used to collect the data, it will be an uphill task that limits collaboration, hinders communication, and slows down responses from the IT operations team.

In this chapter we discussed the conflict that exists between the innovation requirements and the support requirements and how AI can transform the traditional IT operations approach to a highly responsive and integrated AIOps. In the next chapter, we discuss an interesting concept—algorithmic companies and decentralized autonomous organization (DAO).

References

1. https://www.gartner.com/smarterwithgartner/
   12-steps-to-excellence-in-artificial-
   intelligence-for-it-operations-infographic/

2. https://www.forbes.com/sites/janakirammsv/
   2017/07/16/artificial-intelligence-is-set-to-
   change-the-face-of-it-operations/#6f0100a11d21

3. https://www.splunk.com/en_us/solutions/
   solution-areas/it-operations-management/
   aiops-platform.html

4. https://readwrite.com/2017/05/15/
   artificial-intelligence-transform-devops-dl1/

5. https://home.kpmg.com/uk/en/home/
   insights/2018/02/advantage-digital-how-ai-is-
   transforming-performance-in-operations.html

6. http://searchitoperations.techtarget.com/
   essentialguide/AI-in-IT-tools-promises-
   better-faster-stronger-ops

7. https://logz.io/product/cognitive-insights/

8. https://logz.io/product/cognitive-insights/

Decentralized Autonomous Organizations = Blockchain + AI + IoT

Blockchain, AI, and IoT are all technology disruptions: Blockchain promotes decentralized applications in an open-data environment, AI provides intelligence through a centralized data and analytics platform, and IoT enables devices/ machines to be connected via a centralized cloud storage and processing service.

In the previous chapters, we discussed AI and its influence and the potentially explosive value generation if combined with IoT, RPA, and IT operations. In this chapter, we discuss another aspect of AI in a highly decentralized environment

© Soumendra Mohanty, Sachin Vyas 2018
S. Mohanty and S. Vyas, *How to Compete in the Age of Artificial Intelligence*,
https://doi.org/10.1007/978-1-4842-3808-0_9

and how, if combined with blockchain technology and IoT, it can deliver far-reaching and amplified outcomes.

First, let's have a quick overview of blockchain.

Blockchain Primer for Managers

We have witnessed two waves where initially scarce and expensive resources became cheap and technologies then emerged to exploit these resources. When computation (transistors) became cheap, PCs (personal computers) emerged; when bandwidth became cheap, the Internet emerged. Similarly, blockchain is the third wave where cheap storage allowed us to create a distributed (distributed ledgers to record ownership), open to all (yet secure) way of recording important information, thus introducing a robust, transparent, and trustworthy mechanism to issue and transfer assets in the virtual world.

Blockchain (distributed ledger) is a secure, distributed, and immutable database, accessed by all participants in a distributed network where transaction data is recorded (either *on-chain* for basic information or *off-chain* in the case of attachments) and audited. The data is stored in structures called *blocks*. The blocks are connected to each other in a *chain* through a *hash* (each block also includes a *timestamp* and a link to the previous block via its *hash*). The blocks have a header, which contains metadata information and the content (real transaction data). Since every block is connected to the previous one, as the number of participants and blocks grow, it becomes extremely difficult to modify information in the block chain without obtaining the network consensus.

Blockchain has three major characteristics:

- *Decentralized control*: Blockchains operate on a shared control framework wherein it offers a common mechanism for participants for a common cause. No single entity in a blockchain network has exclusive rights granted to it that others do not have. This allows for the smooth formation of decentralized networks for various transactional services. This is highly influential in cases where best practices need to be made aware to everyone so as to improve the process as a whole, make it beneficial to the end user, and make it easy for all to ensure better governance and compliance.

- *Assets exchange*: This is perhaps the most important feature of blockchain, since all transactions are recorded and no single entity has total control over what happens in the network. The system allows for transparent exchange of value items, which are data units that have value, like digital currencies, energy unit data, stock unit data, land deeds, educational certificates, and much more. It becomes easier to maintain in a trusted network critical information exchange and can revolutionize the way in which transactions are carried out with blockchain technology compared to traditional mechanisms. More reliability, security, and transparent control will ensure that transactions are viable under all circumstances and do not violate ethical policies.

- *Immutability*: A transaction, once it's recorded, is permanent and can't be erased from the network. There is a permanent digital audit trail that lets everyone in the network see who did what. This forms the basis for smart contract management when contractual obligations become easier to validate from all parties involved. This will ideally mean complete elimination of fraudulent practices and keeping a check on all transactions carried out on such networks. It will become impossible for someone or a group of people to influence the entire network to commit a faulty transaction. No matter how long the network exists, all previous actions would be available to monitor and inspect for all members and stakeholders involved.

There are two methods applied to validate the transactions in the network.

Proof-of-work asks the participants (called *miners*) to solve complex mathematical problems in order to add a block. When you want to set a transaction, this is what happens behind the scenes:

- Transactions are bundled together into what we call a block.

- Miners verify that transactions within each block are legitimate. To do so, miners solve a mathematical puzzle known as proof-of-work problem.

- A reward is given to the first miner who solves each blocks problem.

- Verified transactions are stored in the public blockchain.

Proof-of-stake instead tries to attribute more mining power to participants who own more stakes (coins). It is still an algorithm, and the purpose is the same as of the proof of work, but the process to reach the goal is quite different. Under this system, forgers (the PoW equivalent of a miner) build blocks based on their stake in the blockchain's network.

Blockchains follow two approaches to grant access to participants in the network:

- *Permission-less blockchains*: Anyone can join the network and participate in the process of block verification to create consensus and also create smart contracts. Bitcoin and Ethereum blockchains are a few examples, where any user can join the network and start mining.

- *Permissioned blockchain*: Only a restricted set of participants have the right to validate the block transactions. A permissioned blockchain may also restrict access to approved participants for creating smart contracts. Examples: R3 (Banks), EWF (Energy), and B3i (Insurance).

What happens when we merge IoT and the blockchain into a single, powerful platform?

Blockchain and the Internet of Things (IoT)

In principle, it makes a lot of sense, and there are several clear advantages of the idea of machines communicating with each other and humans and operating in a secure, trusted way via blockchain. IoT is all about continuously providing information about the state of the machines, Blockchain is all about an encrypted, distributed transaction recording system. Put them together and in theory, you get a verifiable, secure, and permanent method of recording data generated by smart machines.

What are the real-world benefits of blockchain and IoT convergence?

- *Solving the issue of oversight*: In a connected world where there are multiple networks owned and administered by multiple organizations and vendors, it is important to have the oversight of what is happening across the entire network and when. In this type of distributed operations scenario, a permanent immutable record means custodianship and traceability. In the real world, physical goods passing between points in the supply chain is a classic example of loss of oversight. IoT devices continuously transmit data and blockchains register every single change to the state of goods in transit, which are visible to anyone authorized to connect to the network. This solves the problem of lack of oversight. If something goes wrong—breakages occur, theft, pilferage, etc.—then the blockchain record would make it easy to identify the weak link and raise alerts for action.

- *Establishing trust*: The use of encryption and distributed immutable storage means that the information recorded in the blockchains at any point in time can be trusted by all parties involved, including machine-to-machine interactions. The "smart contract" facility enabled by blockchain networks helps establish agreements that will be executed when conditions are met. This is highly useful when it comes to, for example, authorizing machine-to-machine interactions and task executions when conditions indicate that tasks in the previous step are completed. IoT-enabled devices will securely interact and exchange information without any human involvement.

- *Ensuring heightened security*: Both in the industrial world and consumer world, much of the data generated by IoT is highly sensitive and mission critical. This data needs to be shared with other machines and services in order to be useful. But it also means that this data is vulnerable and open for hackers to potentially misuse. This is where blockchains bring in robust access control, encryption algorithms, and security mechanisms that are difficult to breach.

What happens when we merge AI and the blockchain into a single, powerful offering?

Blockchain and AI

There is a bit of AI in everything around us, be it searching for something on Google to buying something through Amazon, commuting to work, playing our favorite playlist, monitoring health conditions, ordering groceries, etc. Our decisions are being influenced by algorithms that have been developed to cater to almost any kind of human persona and needs.

Blockchain, on the other hand, provides a way to exchange value-embedded data without friction and with trust, thereby making the data secure and ensuring that it is used for the intended and previously agreed upon purpose—nothing more, nothing less.

At a broad level, there are several ways in which AI and blockchain can complement each other.

AI Needs Encryption

Data on a blockchain is highly secure owing to the cryptography technology record transactions on blockchain. Blockchains are ideal for storing highly sensitive data through private keys in order for all of the data on the chain to be secure.

There is also a larger need for AI algorithms to work on the data, including the processes while it is still in an encrypted state, because any phase of the process in which the data is unencrypted represents a security risk.

Blockchain Can Help AI Become Explainable

AI algorithms analyze a large number of variables to "learn" patterns hidden within a vast amount of data and arrive at predictions. The learning process and the complexity of algorithms often becomes incomprehensible to end users—why the algorithm recommended what it recommended, how the algorithm arrived at the recommendation, which variables it used and which ones got dropped and why, and many more such questions.

AI offers huge advantages in many fields; however, if the outcomes are not explainable, then they won't be trusted by the end user. Blockchain technology can help AI record the entire learning and decision-making process, thus providing a level of transparency and insight to the end users.

AI Can Make Blockchains More Efficient

Due to heavy focus on encryption, blockchains require large amounts of compute and processing power. For example, hashing algorithms to mine blocks on the blockchain effectively try every combination of characters until they find one that can verify a transaction. Sophisticated machine-learning algorithms can optimize the blockchain mining algorithms, if they are fed with the right training data.

AI Can Make Blockchain Scalable

Blockchain by design is immutable, which means managing growing volumes of data becomes a stiff challenge. "Blockchain pruning" (removing data about inactive or closed down transactions) could be a possible solution. AI algorithms can play a role in automatically archiving using data sharding techniques to make blockchain more efficient and scalable.

Integrated AI and Blockchain Value Propositions

Improved Business Data Models

While existing AI-driven business models rely on data produced by organizations and their ecosystem partners, there are drawbacks when it comes to transparent data sharing. Privacy concerns, abuse of data, and fraudulent contracts make it impossible to create fully open data systems. For AI to produce the most accurate decisions, it needs virtually no barriers to information access from the entire business ecosystem. This frictionless information access would be provided by blockchain technology, as all stakeholders would be more willing to share information since no single entity can take ownership of the network.

Data sources from all key players can easily be integrated with a common AI-powered analytics platform. The way the data is going to be used would be defined in the blockchain ledger for the whole transaction, so there is no chance of data owners being subjected to scenarios where they would lose the credibility of their data or their data would lose its value.

Once an uninterrupted supply of data from multiple stakeholders is established, AI systems can dive deep into them and study the patterns and other aspects to uncover insights that were never witnessed before. As the number of participants in the network grows, there would be more possibilities to extract more genuine insights and the systems can train themselves to better respond to scenarios based on how each stakeholder would respond, by identifying best practices and discovering the best solution to a new scenario.

Newer Insights to Discovery

With transparent information access, data silos could be more efficiently merged and this would help AI to maneuver through newer data combinations and discover new patterns. With blockchains it is easier to eliminate data manipulation, so AI can be used to create new authentic classifiers and filters for data, since it is possible to verify their authenticity on a decentralized blockchain framework.

This is particularly important in scenarios that do profiling and segmentations for better predictive customer engagement. More data sets would allow micro-segmentation and hyper-personalization.

Intelligent Predictions

AI has come a long way in helping us predict outcomes based on data collected from various sources. However, at times such predictions may become factually incorrect because of faulty data generation systems, fraudulent tampering of data sources, or loosely governed analytics methodology employed in AI systems. With blockchain it is easy to authenticate data sources and the analytical methods to be applied to the data, as they would have to agree to predetermined contracts and best practices. This paves the way for AI systems to work only on authentic and genuine data sets, thereby resulting in accurate predictions.

More genuine data sources would ensure that AI systems work more efficiently to mine insights. These insights would then be utilized by deep learning algorithms to arrive at more factually correct decisions. This would ensure better predictions of end consumer behavior for businesses or end user behavior for better governance processes.

Data Intellectual Property Rights

AI-enabled data models always create great success stories that inspire others. But very often creators of such models refuse to share their data model information for lack of protection and copyright policies. Even though there exist copyright policies, data models can be masked intelligently to showcase totally different architectures that can't be identified even by the creators of the original data models. However, with the integration of blockchain technology, creators can easily share their data models without losing exclusive rights and patents on their discoveries. A tamper-resistant global registry can be maintained with your cryptographic digital signature that contains both your data and its associated models. Who does what and when with this data would always be available on the public domain, thereby making it virtually impossible for anyone to claim authorship of your creations.

With integrated AI tools, it is possible to analyze trends and behavior from these data models and enable powerful insights discovery from them. It also empowers businesses and content or data owners to claim their due privileges and rights to intellectual property for time immemorial. It offers an attractive way to monetize original data or content for professionals in multiple streams.

Autonomous Organizations

This is virtually the dream scenario of AI. An environment where machines perform tasks without human interventions. All that is needed is a globally verified working policy and set of instructions that need to be secure, tamperproof, and not owned by any single entity to set up a training and learning environment for AI-enabled systems. With blockchains, this is easier done than said and before you know it, AI systems can read data, process it, and operate your key business tasks. Chances of failure or faulty operational outcomes are negligible since the entire system is governed by smart contracts residing on decentralized blockchain networks. The network nodes store states, parameters, behaviors, results, etc., of the entire operations systems and allow AI systems to act accordingly to predetermined smart policies.

Such a level of autonomous operations has never been witnessed before; however, with AI-enabled data secure blockchain systems, this mode of autonomous operation is quite feasible. Deviations from predetermined behavior only occur when the system is asked to deviate by controlling nodes under the supervision of all stakeholders. Such a secure and transparent monitoring and control mechanism ensures that things do not go haywire at any point in time and human intervention is kept at a minimum, or in ideal conditions there will not be any involvement at all.

A few examples of AI and blockchain technologies coming together to deliver industry specific outcomes are discussed in the following sections.

Smarter Finance

AI and Blockchain technology are already having a significant impact on financial services on an individual basis. Their combination would result in even better transformative ways for financial entities to conduct transactional services. For example, forming banking consortiums to derive new service layers, policies to prevent fraudulent usage and access of customer information, etc. can all be achieved with AI and blockchain combined.

While the insights would be offered by AI, the underlying trusting platform would be blockchain. Banks would willingly contribute their data and information for a greater cause, as there is no single entity that has the ownership of the entire consortium. With such a secure and connected sharing ecosystem, AI systems can traverse a wide variety of demographic and market data, thereby uncovering new insights about business growth and potential opportunities to serve customers better.

Intelligent Retail

E-commerce and digital retail solutions for in-store sales have revolutionized consumer shopping experiences like never before. Today one of the biggest investments made by retail organizations on the technology infrastructure side are analytics and prediction systems. For a modern day retailer, getting to know a consumer's buying decision before he or she actually makes purchases is the USP for successful experiences and checkout.

AI-enabled systems are already driving more sales for retailers with more intelligent prediction capabilities. However, there are still areas where AI is left to work on unverified data sets to arrive at decisions. Data about products or component quality from partners and or vendors is often not verifiable. If you provide the most suitable product but it ends up being a failure in terms of quality, then the entire retail experience is destroyed at that moment and could result in the customer leaving you forever.

With blockchain-integrated systems, it is possible to verify claims of vendors and suppliers regarding the quality and specifications of the product, as they would be available for inspection and analysis at any point in time. Accountability of all stakeholders increases and this collective responsibility results in only the best products reaching the shelf or on display in an online store. With more accurate data, AI systems can uncover new insights like, for example, new pricing points based on authenticity of materials (sourced from prime locations or exquisite places), and this leads to more value for creative marketing as well.

Hassle-Free Customer Loyalty Programs

Almost every business has a loyalty program (part of their customer relationship and marketing strategy), through which they reward their valued customers and encourage them to stay loyal (through differentiated treatment, disproportionate discounts, access to yet to be generally available products or services, coupons, vouchers, etc.). A good loyalty program is not only about rewarding valued customers, rather it's an important strategy to strengthen customer relationships.

In reality, from a consumer perspective, we are subjected to multiple loyalty programs, each having its own rewards mechanism, maze of point systems, and cumbersome redemption processes. The result is that the consumer is not benefiting and the business is not enjoying customer loyalty in the true sense.

Blockchain may just be the answer. For consumers getting lost in the multitude of loyalty programs (physical cards or digital wallets), blockchain could provide a single unified platform that seamlessly manages each loyalty program option, including the limitations and redemption rules. How?

Blockchain enables a distributed ledger of transactions shared across a network of participants (consumers as well as businesses). When a new transaction occurs (a loyalty point is issued, redeemed, or exchanged), blockchain will issue an algorithm-generated token against that transaction. The tokens are grouped into blocks and distributed across the network, updating every ledger at once. In addition, blockchain ensures the blocks are validated and linked to older blocks, thus creating a secure and transparent audit trail of all transactions, without the need for intermediaries.

Customer loyalty programs on blockchain would mean a robust and trusted partner network for businesses, where they can collaborate and offer more options outside of their own core products and services, thus breaking out of the narrowly defined individual loyalty programs and reducing customer hassles.

Creative Passport

These are smart contracts for music industry. A major painpoint in the music industry is that the songwriters, producers, and musicians are always in the blind spot as far as how the copyrights are handled, how their royalty payments are calculated, and in general how digital products are managed and distributed and how the consumers are consuming their digital products.

This poses significant challenges to both the producers and the consumers.

Blockchain, with its capability to provide a verified, transparent, and distributed global registry of music for managing the digital product's lifecycle, can eliminate the current creative industry challenges and at the same time provide a seamless experience for anyone involved in creating or interacting with music. For example, the activity of listening to a song would automatically trigger a chain of events in the blockchain system for everyone involved in the journey of that song with anyone who wants to experience the song or do business with it (an individual consumer, a digital service provider like Spotify or iTunes, a radio station playing the song, or a film production crew using the song, etc.). The blockchain system will have the added advantage of storing metadata about the music, sort of a "creative passport," which could then be updated and accessible to everyone.

Transparent Governance

Every year, governments across the world spend billions on social improvement and citizen welfare schemes. However, due to corruption and inefficient middle layers and lack of proper auditing systems, a good portion of that money does not reach its intended recipients. Besides this, there is a huge waste of manpower involved in performing redundant and mundane tasks such as data entry, proofreading, verifications, log entries, etc.

By bringing in smart automated management systems powered by AI and built on blockchain technology, a high degree of efficiency and transparency can be achieved in governance mechanisms. Using AI, it is possible to free human resources for more important activities where all redundant tasks can be automated. A great deal of insight generation to see progress of the schemes and policy implementations can be created by AI systems that eliminate any form of human interference or tampering, thereby avoiding corruption.

With Blockchain, it is possible for citizens and governments to establish trustworthy policy implementation and data exchange mechanisms without worrying about faults and corrupt or malicious involvement. Citizens can verify claims of government spending and escalate concerns with proof. It is possible for governments to monitor all activities of all departments, thanks to the smart audit trails available in blockchain systems.

Globalized Verification Systems

Today there are a lot of verification systems across the globe that require information sharing and integrated view crossing boundaries, be it citizenship and identity verification, immigration details, credit and financial history, etc. This data is highly sensitive and can be manipulated if it lands in the wrong hands. However, for the smooth functioning of global systems, such as immigration, banking, educational, scientific research programs, etc., it is vital to have transparent information access for verification of claims and identity.

AI-enabled systems would help to study deep patterns stored in data about people and predict their credibility scores or financial health or identity, which is vital for security reasons such as in preventive measures for terrorism-related offences or financial crimes such as anti-money laundering, etc.

A globally decentralized registry for crime records, financial fraudulence, etc. would ensure that AI systems can train themselves of verified data to arrive at conclusions and spot outliers or anomalies. Blockchain-powered global verification systems would help to bridge the gap between global investigative and verification agencies, thereby smoothening verification processes across the globe and ultimately resulting in better cross-border commutes.

Innovative Audits and Compliance Systems

With blockchain creating a decentralized network for independent verification of claims, it is easier for authorities as well as businesses to adhere to compliance and regulatory requirements. Every transaction on a blockchain network leaves a digital trail that cannot be tampered with and is irreversible. This trail can be audited by any authority and can be used to verify claims of authenticity.

Using AI, data from these audits can be utilized to provide better services to customers and citizens, such as accurate pricing and legal information about real estate, tax calculations, and tax accountability, compliance to nationally or globally accepted quality standards for services and products, and much more. The possibilities are limitless as more authoritative data can help AI systems provide uninterrupted and factually correct information and services to end users.

Data and AI Monetization

Monetizing collected data (our data) is a huge revenue source for large companies, such as Facebook and Google.

With everything we do on the web or using apps, we leave behind huge data footprints. This data is aggregated and eventually used to cross-sell and up-sell stuff to us. This makes data incredibly valuable—more valuable than we the original data generators realize—and we are giving it away for free.

Blockchain enables microtransactions that make it possible for us to own, control, and monetize our own data. The same goes for AI. If we develop AI algorithms using our own data, we can put the algorithms on an AI marketplace for others to use. We are then paid usage based fees.

The Case for Decentralized AI

Until recently, the contemporary AI industry was based on a centralized development and usage model where machine learning solutions were designed, developed, and accessed as a part of cloud-based AI platform and associated APIs. Now we are moving toward the next frontier—decentralized AIs that can run and train on local devices or make decisions in decentralized networks like blockchain.

The existing AI market is increasingly controlled by tech giants like Google, IBM, and Microsoft, all of which offer cloud-based AI solutions and APIs. This model assumes little control of users over the AI products, and in the long run, such a centralized model could lead to the monopolization of the AI market. This could cause unfair pricing, a lack of transparency, interoperability, and limited participation of smaller companies in AI innovation. Fortunately, we are witnessing the emergence of a decentralized AI market, born at the intersection of blockchain, on-device AI, and edge computing/IoT.

The transition to decentralized AI is enabled by new technologies, such as Google's federated learning, that allow for crowd-training of ML algorithms, device-centric AI that run and train ML models on mobile devices, and the use of AI in DAOs (decentralized autonomous organizations) on blockchain networks.

AI and Decentralized Autonomous Organizations

DAOs manifest when we entrust some or all decision-making responsibilities to AI agents. For example, If you are a holder of ownership rights, you can cede your decision making (e.g., yes/no votes) to an AI agent (another smart contract) that will make all the decisions on your behalf. Or, in a more radical scenario, we can entrust AI agents to play the role of managers and take all organizational and business decisions. For example, imagine an AI agent for marketing, where the AI agent selects the best ad companies to place your ads with. After each marketing cycle, the AI agent would assess the ROI and adjust the marketing strategy automatically.

In essence, DAOs take us to a qualitatively new economic reality where AI agents become business administrators and are overseeing business performance and continuously learning from and with other AI agents in a decentralized network. Terra0 is a project proposed by Paul Seidler and Paul Kolling from the University of Arts, Berlin. The concept involves augmented intelligence agents managing and monetizing a self-owned forest. Forestland ownership is structured as an DAO with smart contracts on the Ethereum blockchain. Then, using drones and satellites, the AI agents evaluate the woodstock and decide how much and when to sell in the market. Once the project is up and running, the DAO can pay out debts to its initial owners and eventually turn the forest into an autonomous, self-owned entity that controls its own resources.

Recent advances in decentralized AI have been made thanks to on-device optimization of AI/ML for smart phones and production of dedicated chips for mobile AI and for desktops (e.g., Google's TPU).

Decentralized AI gained powerful momentum in April 2017 after Google announced its new Federated Learning concept. This innovation signals a transition to fully decentralized learning and device-centric AI where machine learning models are trained directly on user smart phones. Keeping the privacy of user data intact, Google can now outsource AI training to Android users, enabling on-device improvement of shared models. Federated Learning will solve the problem of high-latency and low-throughput connections where users have to connect to remote servers to use ML software.

The move toward device-centric AI can also be seen in the release of Google's TensorFlow Lite, a mobile version of a machine learning library fined-tuned to the computational and power constraints of smart phones. In June 2017, Apple followed Google's lead by releasing its Core ML library for iOS devices. The library ships with the optimized general-purpose ML models and tools to convert third-party models into the iOS format. Making models available locally without a network connection will make it easier to develop mobile applications with AI functionality.

In the long run, a combination of on-device AI and decentralized learning will make AI more democratic and widespread than ever before, resulting in more and more organizations becoming DAOs.

What is accelerating this trend?

- *Autonomous AI on blockchain*: DAOs are algorithmic companies run by autonomous AI agents. When you put these autonomous AI agents on an Ethereum blockchain, you can effectively manage distribution of royalties, subscription payments, smart contracts, and more.

- *Decentralized learning at the edge*: On the device, AI capabilities such as mobile machine learning libraries from Apple's CoreML allows complex AI algorithms to run on IoT devices like sensors, security cameras, drones, or autonomous vehicles.

- *Decentralized intelligence networks*: Decentralized networks built on the blockchain such as SingularityNET allow any company or researcher to monetize their AI solutions and get access to a variety of AI algorithms. SingularityNET enables cross-AI capabilities through protocols that support data exchange and sharing across different algorithms, which is helpful in building multitier and connected AI applications.

Such applications can combine multiple algorithms and perform different sub-tasks and then have access to the training data exchanged on the network. One example of such an approach is the development of a comprehensive AI-based cybersecurity solution. Currently, there is no single software package that handles all security-related tasks, which means that companies have to use various centralized AI solutions and customize them to their needs. Developing such a solution also brings about the problem of data security and domain-specific knowledge.

Homomorphic encryption that makes different data sets visible to different classes of users in different aspects can solve this problem. Companies can also combine the expertise of different cybersecurity AI agents on the network, which will safely exchange security information, outsource tasks, and cooperate in solving common security issues. Such decentralized network will involve efficient division of labor and offer access to solutions without having to manually obtain data and customize algorithms.

Decentralized intelligence has a number of advantages over centralized solutions in the following scenarios:

- *You need an autonomous AI solution that runs in the decentralized environment and implements contractual obligations:* By definition, centralized proprietary solutions cannot be exposed to many users in the decentralized network. If your goal is to run a fully autonomous AI agent that makes smart managerial decisions and distributing profits, decentralized AI on blockchain is the way to go.

- *You need an AI optimized for the on-device performance and not dependent on network connectivity:* Due to network connectivity problems, battery power constraints, and low computing power, mobile devices are not a good option for running cloud-based AI software. In particular, high latency and low throughput can compromise the speed and performance of AI applications, adversely affecting user experience. Whenever you need to run and train AI on mobile devices, a decentralized network becomes a better option.

- *You want to sell your AI algorithms while maintaining proprietary rights:* There is no way to sell your AI algorithm and retain proprietary rights for it (the same goes for selling mobile apps in the app store right now). To fill this gap, decentralized AI networks offer an opportunity for developers to make their algorithms available for commercial usage as AI as a service.

At the same time, centralized AI still remains a good option if you need a very generalized ML model that you can easily plug into your application. Google, Microsoft, and IBM have developed the best generalized machine learning models on the market that are trained on huge data sets and built according to the top ML standards and bleeding-edge ML algorithms. Reinventing the wheel is not an option if you want proven image or speech recognition features in your application. A more viable solution is to use cloud-based ML APIs provided within a pay-as-you-go model, which ensures cost efficiency and scalability of your AI-based solutions. Major providers of centralized AI have a comprehensive suite of services for image and video recognition, emotion AI, speech recognition, predictive modeling, and other common AI/ML tasks.

In the long run, decentralized solutions can produce the radical democratization of the AI market, optimization of solutions for a wide variety of use cases, easy integration and communication between different algorithms through a single protocol, and the development of interoperability standards, which will ultimately lead us to the era of AGI (artificial general intelligence).

Conclusion

The combination of the blockchain technology and artificial intelligence is still a largely undiscovered area. Traditional AI methods follow a centralized development and deployment pattern. You have a cloud-based AI platform, you ingest data and do your predictive models and then, through APIs, access the models from various business processes or applications. There are a number of challenges to this approach: latency, data security, network bandwidth, performance, etc.

Now, imagine a highly decentralized approach where your predictive models run, train, and even make decisions on local devices in decentralized networks like the blockchain. This is decentralized AI!

There are significant benefits of decentralized AI over traditional AI:

- Minimal latency (no dependency on a network connection)

- Training is more efficient (done in a decentralized way)

- Less power consumption (again, no dependency on a network connection)

AI can be incredibly disruptive and must be designed with utmost precautions. Blockchain can greatly assist with the cause. How the convergence between these two technologies will progress is anyone's guess.

In this chapter we touched upon several disruptive technologies (blockchain, AI, IoT, and cloud), each extremely powerful on its own. When they are combined, they can deliver far-reaching transformative outcomes. In the next chapter, we touch upon the thorny topic of ethics and AI, which has raised enormous debates worldwide.

References

1. https://www.forbes.com/sites/forbestechcouncil/2017/11/16/why-decentralized-artificial-intelligence-will-reinvent-the-industry-as-we-know-it/#3bcedba7511a

2. https://blog.bigchaindb.com/blockchains-for-artificial-intelligence-ec63b0284984

3. https://www.forbes.com/sites/forbestechcouncil/2017/11/16/why-decentralized-artificial-intelligence-will-reinvent-the-industry-as-we-know-it/#3bcedba7511a

4. https://blog.bigchaindb.com/blockchains-for-artificial-intelligence-ec63b0284984

5. https://blog.bigchaindb.com/blockchains-for-artificial-intelligence-ec63b0284984

Ethics and AI

AI, especially ANI (Artificial Narrow Intelligence), is beginning to occupy a unique place in our lives. There are virtual personal assistants, hosts of intelligent apps on our smart phones, self-driving cars, intelligent home appliances, smart buildings, smart homes, and many more. No other technology has the kind of far-reaching implications that AI has, across so many industries and all spheres of our lives. This is what makes it so fascinating; it is truly incredible and no less than a sci-fi plotline.

For a technology designed to mimic human-like intelligence, understanding the darker side of AI is also equally important and probably requires defining the dos and don'ts at a policy level. There are many questions we need to address, if we want to truly leverage the transformative power of AI.

In the next 20 years, experts predict that AI will continue to make great strides in transforming and in many cases eliminating hosts of human tasks. If this innovation is done in an ethical way and keeping human-machine collaboration as the goal, we can build a future in which humans are not competing with machines. Instead, we will be entering into a new era of jobs that requires no manual effort but higher emotional intelligence.

It is perhaps a way of life now to hail a taxi at the tap of a smart phone screen. Imagine the same thing, extended to getting a self-driven taxi! Imagine the extent of woes that the traffic management officials will have to go through to manage transportation and safety to accommodate the next wave of driverless cars. For sure, the department of transportation has their work cut out for them; they will have to innovate too.

© Soumendra Mohanty, Sachin Vyas 2018
S. Mohanty and S. Vyas, *How to Compete in the Age of Artificial Intelligence*,
https://doi.org/10.1007/978-1-4842-3808-0_10

This is just one of the example that shows how the implications of AI are unprecedented. As we continue to innovate and solve new problems using AI, we must simultaneously pay attention to ethical usage and policy related aspects because there are implications to the society at large, not just a few industries or a few individuals. We must take a democratic approach to the future of technology led disruptions.

Ethical Issues in Artificial Intelligence

AI, in many ways, is pervasive and provocative at the same time. We need guidance and frameworks to ensure ethical usage of AI, and this is the right time in the AI evolution phase to consider ethics in relation to AI seriously.

The ability to be introspective has made us what we are. However, it seems that we are in a hurry to outsource this to algorithms. There are grave consequences for such an approach. We are pleased to see how automation is improving the quality of our lives, but we have not sat down and created a list of activities that we would never delegate. We have not even crafted guiding principles for clear demarcation of responsibilities between humans and machines. These things are important. In the absence of values-based standards and guidelines, the biases of AI manufacturers will take over and dictate the terms and conditions.

Ethics is not about saying, no you can't do this. The point is to ask what are the goals you are trying to achieve and how can you attain these goals without infringing on the cultural values we hold dear. You can do things that are entirely legal yet highly unethical. The Socratic process of always asking for contrarian views is not always fun, but basking in the glory of AI achievements and not paying attention to ethical standards is no longer an acceptable stance.

The following sections contain a list of topics that are subject to raging debates in various forums. The camps are still divided.

Unemployment

What happens when all manual jobs are replaced by AI-assisted machines?

Cultural aspects such as position in society and consequently the division of labor are governed by the magnitude of automation. In the pre-industrial age, one's ability to do physical work determined the paycheck. Gradually we evolved and invented ways to automate jobs, not eliminating the physical work completely, but augmenting human skills with machines so that we could do more with less. The result? Newer jobs like machine operators came into existence and slowly the skills orientation moved from physical labor to the cognitive labor, putting more and more emphasis on judgment-related skills.

For example, trucking as an industry currently employs millions of individuals across the globe. What will happen to trucking jobs if self-driving trucks become a reality in the coming years? On the one hand, self-driving trucks might lower risk of accidents, won't show signs of fatigue, will cover the distance in a predictable way all the time, reduce the cost of operations, and deliver significant efficiency gains. Hence, self-driving trucks seem like an ethical choice. The same scenario could happen to the majority of the workforces across other industries and sectors.

The ethical question before us is, when more and more jobs are automated, what will we do with all the time on our hands? Our current employment contracts are based on one fundamental factor—we sell our time to earn enough to sustain ourselves and our families. So, if the prospect of time-based compensation goes away, we need to find newer ways to earn money.

Inequality

How can we reward the machines?

The majority of our current compensation frameworks are based on hourly wages. There are few notable changes where risk and reward mechanisms are in place, but in general our services always equate to a rate per hour. If AI is going to do most of our work at a fraction of a cost, companies will naturally drift toward a newer workforce mix (an increasing number of AI agents and a smaller human workforce). This means fewer humans will get compensated. However, the company owners and others who invested in the company will take home a major share of the revenues earned.

If you have not noticed, this is already happening and it's creating a widening wealth gap. Companies whose business model revolves around this algorithmic economy are employing fewer humans and sharing the wealth among the workforce from the economic surplus they are creating.

The ethical question before us is, if we're truly moving toward a no manual-job society, how do we structure a fair compensation mechanism for the workforce consisting of fewer humans and more machines?

Humanity

How will machines affect human behavior?

On the one hand, we are relishing the fact that we are letting machines think and act on our behalf. On the other hand, we are truly altering human behavior. Here are few examples that illustrate how we are being influenced:

- Websites are designed by taking into the account the minutest levels of detail that would appeal to an individual's liking.

- Recommendation engines push additional products to us by suggesting "people like you have bought these other things".

- While you are driving past a supermall, your phone vibrates with a discount just for you.

- Intelligent apps on your smart phone suggest which route to take to get to your destination faster.

The right information at the right time through the right channel is all good, but the side effect is that we have stopped thinking about anything anymore.

The ethical question before us is this—first, by allowing AI agents to think and act on our behalf, are we progressing toward a world where we are becoming increasingly uneducated about our surroundings? Second, if the AI agents are becoming better and better at modeling human behavior, is there a possibility that the same AI agents will be used to direct human attention and trigger certain actions that are detrimental to the very existence of the human race?

Artificial Stupidity

How do we ensure that machines do not become biased?

Humans develop cognitive capabilities by learning from their environment; machines also go through a similar learning phase to acquire intelligence. Humans learn bad things if they are exposed to bad environments; machine learning goes through a similar risk if the data is incomplete or purposefully distorted.

Machines need to learn continuously and need to be exposed to a wide variety of data sets in order to be prepared to handle real-world occurrences. Just relying on training data sets that are curated based on the data that's available won't suffice if we want machines to take on real-world challenges.

Humans are not always fair and neutral. Machines also can exhibit similar unfair and irrational behavior. AI systems are created by humans, thus there is a high likelihood that humans will introduce judgmental bias into the very machines they build. Bias can also creep into machine behavior in many different ways—data bias and design bias are the most prominent ones.

The ethical challenge before us is, if our future is going to be completely dependent on AI systems, we need to ensure that the machines perform as expected and aren't biased.

Security

How do we keep AI safe from evil intentions?

The more powerful a technology becomes, the more possibilities open up for malicious intent. Autonomous systems require a greater responsibility of making them secure, and it is not just adversaries we need to worry about. What if AI agents become so focused on achieving their goals, that they recommend and implement things that may bring disastrous consequences for us? For example, what if the goal of an AI system is to find solutions for cancer, and after careful considerations of numerous diagnosis results, root causes, treatment plans, and effectiveness of medicines, it realizes that the most effective and best way of solving the cancer problem is to kill everybody on the planet? From a machine's point of view, it has found the solution. From human point of view, it is catastrophic.

The ethical question before us is, how do we ensure that there are enough checks and balances in place before we start using AI systems across all spheres of our life and as ubiquitously as we want to?

Singularity

How do we manage artificial super intelligent systems?

Human evolution is almost entirely due to our intelligence and our ability to adapt to changing conditions. However, in our zest to invent more and more artificially super intelligent systems, we may get into a scenario where the machines are the most intelligent beings on earth, far superior than humans! This state is called "singularity".

The ethical question before us is, even if it is far fetching, someday a sufficiently advanced machine may come to life, and this machine will be able to anticipate what we are anticipating, so how do we stay a step ahead?

Machine Rights

How do we define a legal framework for AI?

We have seen how mechanisms of reward and aversion play a decisive factor in human lives. Reinforcement learning in particular applies a virtual risk and reward mechanism to let AI agents learn and become adaptive to the environment.

The ethical question before us is, once machines as entities attain sufficient maturity levels to see, sense, think, and act autonomously, they will demand a legal framework to protect and manage their share of rights. Should the intelligent machines be treated like humans? Should they have a parallel legal and grievances system? When an AI agent is penalized for wrong doing, how do we provide a support system to make it understand its mistakes and stay motivated to do well in the future? How do we resolve situations where the machine is doing all the right things but the human-in-the-loop is in the way?

The Board and CEO's Roles in Ethical AI

AI's predictive measures are meant to provide intended outcomes. Out in the real world, this is the only side that matters. However, just as humans are imperfect, subjective, and prone to corruption, machines are too. In other words, despite established laws and protocols, dubious actors in our society find a way to game the system. Similarly, AI's usefulness is open for people to use it for malicious or self-serving purposes. The stakes are getting higher.

Hence, companies, especially those with access to vast pools of data and that develop and deploy AI products and services, should not only demand more responsible work ethic from engineers, entrepreneurs, and executives, but should also establish more assertive boards and committees to make ethical AI their top strategic priority.

The Board's Role

Companies in traditional industries turn to decades of laws, regulations, and litigation for guidance while defining corporate ethics. However, for something as transformative and disruptive as AI, there is no prior knowledge. Norms and standards are still emerging; laws, regulations, and legal precedent are scarce. That is why it is important for boards to become aggressive proponents of corporate ethics for AI, making it a top priority alongside other concerns such as growth, profitability, M&A targets, and succession planning.

Boards should hold CEOs accountable to making AI ethics an enterprise priority. For example, directors should not take it on face value that the company's AI offerings and initiatives do not inadvertently promote biases. Instead, board members should stipulate that AI offerings and initiatives should be continuously be subjected to stress tests to uncover AI biases and take remedial actions.

The CEO's Role: Governance Excellence on AI Ethics

CEOs can build their AI ethics and governance prudence in three ways.

First, CEOs should strengthen and expand their own knowledge, covering issues ranging from AI bias to best practice in transparency and accountability. They should also be deeply aware of the implications of their company's AI products and services, including data sources and nature/type of algorithms.

CEOs should also set up a dedicated think-tank to collect industry wide information about AI ethics-related litigation and concerns raised by customers and views of policymakers. This will help CEOs conduct management briefings across the enterprise and raise awareness levels.

Secondly, CEOs should include senior advisors into their think-tank who can bring in additional perspectives regarding implications of AI to the boardroom. For example, if your company is offering HR-related AI products and services, it is crucial to bring in a senior advisor to critically review whether biases related to diversity, inclusion, labor, and civil rights regulations are adequately marginalized.

Third, CEOs must realize that ethical AI can't be an afterthought. Enough checks and balances should be built right into the core of AI systems development process to ensure that technologists don't get carried away in developing AI systems without paying attention to making AI transparent. We cannot ignore the risks of deploying AI systems that work exceptionally well on the one hand, but are not fully explainable on the other. Hence, CEOs must ensure that the enterprise is equipped with the right tools, methodologies, and review mechanisms.

The Technologist's Role

Technologists and AI experts must pay attention to the purpose and goals of the AI system. It is easier to say that the algorithm is highly sophisticated but the data is biased. Real-world data will be always biased, because it is situational. It is the responsibility of AI experts to fully understand the context behind the data and then design algorithms to support human values.

Conclusion

It is important to decouple the pursuit of fairness in AI from commercial interests and create room for fair and transparent AI solutions. To that end, making AI explainable is of paramount importance.

Companies building their business strategy based on AI's disruptive power are caught midway: What are the costs and benefits of making their AI transparent? How do they protect their IP and competitive advantage?

Admittedly, there are rules of thumb or best practices right now. Perhaps, for every AI system, if we can detail the kind of data used, details of data sources, parameters used in the prediction model development, accuracy levels achieved, and statistics about false-positive and false-negatives, we may achieve a generally agreeable transparency level that may further lead to establishing a commonly acceptable set of standards for ethical AI.

In this chapter, we discussed several ethical related viewpoints of the consequences of AI. In the next chapter, we bring together all the lessons from previous chapters and try to establish certain approaches to building a human-machine collaborative ecosystem.

References

1. https://80000hours.org/articles/ai-policy-guide/

2. https://medium.com/artificial-intelligence-policy-laws-and-ethics/the-ai-landscape-ea8a8b3c3d5d

3. https://www.cs.ox.ac.uk/efai/category/codes-of-ethics/

4. https://www.wired.com/story/ai-research-is-in-desperate-need-of-an-ethical-watchdog/

5. https://mashable.com/2015/10/03/ethics-artificial-intelligence/#kyYeZZHNMsqD

6. https://techcrunch.com/2017/01/22/ethics-the-next-frontier-for-artificial-intelligence/

7. http://www.zdnet.com/article/artificial-intelligence-legal-ethical-and-policy-issues/

8. https://www.weforum.org/agenda/2016/10/top-10-ethical-issues-in-artificial-intelligence/

9. https://www.weforum.org/agenda/2016/10/top-10-ethical-issues-in-artificial-intelligence/

Putting It All Together: Toward a Human-Machine Collaborative Ecosystem

British botanist Arthur Tansley in the 1930s developed the concept of the ecosystem, inspired by how a localized community of living organisms interact with each other and use the environment to sustain, thrive, and even adapt to changes in the environment.

© Soumendra Mohanty, Sachin Vyas 2018
S. Mohanty and S. Vyas, *How to Compete in the Age of Artificial Intelligence*,
https://doi.org/10.1007/978-1-4842-3808-0_11

In 1993, James Moore extended the concept to modern business, where companies co-evolve capabilities, work cooperatively and competitively to support new products and satisfy customer needs, and eventually fuel the next round of innovations.

Earlier we discussed how the economy has moved beyond large "self-contained" corporations to network based companies, enabled largely by digital technologies, abundance of data, massively increased connectivity, and cloud computing, Apple Inc. explicitly designed its products and services as an ecosystem that would provide customers with a seamless experience; eBay recognized the emphasis it had to place on deliberately building its "sharing ecosystem". Value creation (wealth creation) have taken altogether a different meaning; denser and richer networks where the producers and consumers collaborate to magnify the outcomes for both. For example, who would have thought that platforms to help us connect to our own tribes (Facebook, Twitter, and WhatsApp) would yield significant business values, after all, these are just people connecting with people systems.

Our needs have always driven the means. We were concerned about our wellness, which drove us to find the means like medical practices, physicians, hospitals, pharmacies, etc. Similarly, we wanted to find sources of energy beyond human muscles and trained animals, so we figured out a way to derive energy from fossil fuels, electricity, coal, and nuclear fusion. Innovations like nuclear power, airplanes, automobiles, the personal computer, radio, Internet, etc. disrupted well established business ecosystems and the societal way of functioning, and each of these innovations can be referred to as *black swans*.[1] These breakthroughs were not breakthroughs at the time of their discoveries. They took years in the labs and then got to the mainstream only after they became cheap enough for general consumption. AI strategy alone might mislead just as much as they inform. We think AI technology is another black swan in the immediate horizon and, therefore, if are to truly leverage the transformative power of AI, we have no other option but to establish and leverage a new ecosystem that is based on human-machine symbiosis.

Human Machine Symbiosis

In 2011, IBM Watson won the *Jeopardy* game show. That was a watershed moment not because the machine beat humans at their own game, but because the possibilities opened up our eyes. What followed was a series of striking breakthroughs in AI—image recognition, speech recognition, and many more—all possible through a technique known as *deep learning*.

---

[1]Nassim Nicholas Taleb, *The Black Swan* (Random House, 2007)

On the other hand, machines becoming smarter than humans has also created considerable uncertainty and speculations about our future relationship with machines. Elon Musk described AI as "our biggest existential threat".[2] Stephen Hawking warned that "the development of full artificial intelligence could spell the end of the human race".[3] The philosopher Nick Bostrom, in his book *Superintelligence*,[4] touched upon technological "singularity," at which point machines will surpass the general cognitive abilities of humans.

All of these debates and speculations are based on a tacit assumption that, because machines can see, listen, and act and are continuously learning, they will outperform humans at various tasks (although in narrow domains as of today) and by the same argument they will soon be able to "outthink" us more generally.

The important question is, instead of competing with each other, is there a way we can orchestrate a mechanism whereby machine intelligence and human intelligence complement one another? The ultimate goal would be to build machines that can think like humans and design machines that help humans think better.

Just like any new technology evolution, AI also has gone through quite a few evolutionary phases. It is important to understand how it all started and how the very original intent was redefined many times.

In the summer of 1955, a conference was convened by John McCarthy at Dartmouth University. The conference was attended by the who's who of AI pioneers. Popularly known as "The Dartmouth Conference Declaration," the original proposal stated: "every aspect of learning or any other feature of intelligence can in principle be so precisely described that a machine can be made to simulate it."[5]

General intelligence (also known as the g factor) refers to the existence of a broad mental capacity that enables an individual to perform different cognitive tasks, for example, an individual's performance on one type of cognitive task (say, the ability to draw nature and sceneries) tends to be comparable to the same person's performance on other kinds of cognitive tasks (say, the ability to draw architectural diagrams). This particular general intelligence capability we don't see in today's AI applications. An algorithm designed to play chess would be completely at sea if it were asked to give product recommendations. In short, the g factor of today's AI agents is confined to a narrow type of intelligence.

---

[2]Elon Musk, speech given at MIT Aeronautics and Astronautics department's Centennial Symposium in October 2014.
[3]http://www.bbc.com/news/technology-30290540
[4]Oxford University Press, 2014.
[5]http://www-formal.stanford.edu/jmc/history/dartmouth/dartmouth.html

In addition, terms like "neural networks" and "deep learning" are falsifying the claim that we are on the verge of creating machines that "think like humans do". While the neural net is inspired by the human brain, in reality it is regarded as a generalization of statistical regression models. Similarly, "deep" refers not to psychological depth, but the addition of structure ("hidden layers") that enables a model to capture complex, nonlinear patterns. "Learning" refers to numerically estimating large numbers of model parameters in regression models.

In short, the astounding success of AI so far has been founded on statistical inference, not on an approximation or simulation of what we believe human intelligence to be.

Licklider's Augmentation

Five years after the Dartmouth Conference, the psychologist and computer scientist J. C. R. Licklider articulated a symbiotic relationship between human and computer intelligence. He proposed a complementary relationship: Humans will define the problem statement, set the goals, and play the crucial role of validating the learning. Machines will do the laborious work of sifting through tons of data to generate insights and predictions. This complementary capabilities and orchestrated task execution will be much more effective than a human alone performing the tasks.

This kind of human-machine symbiosis has already crept into our daily life. Familiar examples include:

- Navigating through busy intersections and complex maze of roadways using GPS apps like Waze

- Searching through massive numbers of books or movie choices using menus of personalized recommendations

- Talking to a machine, giving a command, and the machine responding back with facts, figures, weather patterns, restaurants nearby, and so on

In each case, the human specifies the goal and the criteria (such as "take me downtown but avoid busy streets" or "give me movie options that are moderately funny, hopelessly romantic, and with an Asian theme" or "find me a highly rated and moderately priced Italian restaurant within walking distance"). How does it work? An AI algorithm sifts through massive amounts of data to arrive at predictions. The human then evaluates the machine outputs to make a decision. Every time the human chooses or disregards a particular prediction, the AI algorithm makes note of it and feeds that information back into the learning process, thus improving and learning continuously. In no situation is human intelligence overshadowed, rather it is augmented.

It turns out that the human mind is less predictably irrational than originally realized, and AI is less human-like than originally hoped.

The Linda Experiment: Thinking Fast and Thinking Slow

AI pioneer Herbert Simon brought out another facet of AI in the context of human-machine collaboration. He argued that we humans must settle for solutions that "augment" rather than optimize because our ability to work long hours, remember a multitude of things, and reason well are limited. In contrast, machines do not exhibit work fatigue, make consistent decisions always, and can process massive amounts of input with minimal effort. In addition, they can evaluate millions of factors far more accurately than humans can.

Daniel Kahneman, in his book *Thinking Fast and Slow*,[6] highlighted the interesting yet irrational aspects of the human decision making process.

Whenever there are multitude of choices before us, we get into a decision making process, and naturally it may seem that we would evaluate a lot of data, apply certain rational thinking, recall our past experiences, deliberate with others, and then make a decision. This is what Kahneman calls "System 2" thinking, or "thinking slow."

In reality, we don't always resort to such lengthy and slow processes of evaluating anything to arrive at a decision. Rather we typically lean on our own tribal knowledge consisting of a variety of mental rules of thumb (heuristics) to arrive at a decision. Our own decision-making process may seem narratively plausible, but it is often logically dubious. Kahneman calls this "System 1" thinking or "thinking fast," which is famously illustrated by the Linda experiment.

In an experiment with students at top universities, Kahneman and Tversky described a fictional character named Linda. She is very intelligent, majored in philosophy at college, and participated in the feminist movement and anti-nuclear demonstrations. Based on these details about Linda's college days, which is the more plausible scenario involving Linda today?

"Linda is a bank teller. Or Linda is a bank teller who is active in the feminist movement."

---

[6]Farrar, Straus, and Giroux, 2011.

Kahneman and Tversky reported that 87 percent of the students thought the second scenario was more likely, even though a moment's thought reveals that this could not possibly be the case. Feminist bank tellers are a subset of all bank tellers. But adding the additional hint that Linda is still active in the feminist movement influenced narrative coherence and led the students to opt for the (less likely) second scenario.

In essence, the human mind confuses the easily imaginable with the highly probable, by letting emotions cloud judgments, by hypothesizing patterns in random noise, by associating causality to spurious correlations, and by overgeneralizing from personal experience. Many of the heuristics and wisdom we use to make judgments turn out to be systematically biased. Perhaps it won't be highly provocative if we say that our minds need algorithms to de-bias our judgments and surely our eyes need artificial lenses to filter out spurious correlations.

On the other hand, if there is a way to encode the processing logic behind routine tasks, then it is a safe bet that algorithms will outperform humans on the same tasks. However, such algorithms will lack the conceptual understanding and common-sense reasoning needed to evaluate new or novel situations. To illustrate this caution, let us take the example of IBM Watson winning *Jeopardy*. During the contest a question was posed under the category of "US cities": "Its largest airport is named for a World War II hero; its second largest, for a World War II battle." Watson answered "Toronto." Category "US Cities," but answer "Toronto"! (The correct answer is Chicago, by the way.)

This particular example illustrates that certain strengths of human intelligence, like common-sense reasoning, can counterbalance the fundamental limitations of brute-force machine learning.

One of the fascinating things about AI is that it's been hard to predict which parts would be easy or hard. At first, things like playing chess, a mentally challenging activity for humans, will be the hardest for machines, but it turned out to be easy. On the other hand, things like recognizing objects or picking them up, a fairly simple task for humans, are much harder for machines to do.

In summary, humans need machines to avoid "System 1" decision making traps, and machines need humans to avoid "System 2" decision making traps. Together, they imply that the case for human-computer symbiosis is stronger than ever.

What we need is a human-machine integrated strategy.

The Human-Machine Integrated Strategy (HMIS)

Although the enhancements in AI are making life easier for human beings day by day, there is constant fear that AI-based systems will pose a threat to humanity. People in the AI community have a diverse set of opinions regarding the pros and cons of AI mimicking human behavior.

For example, a neural network trained to detect digits from the MNIST data set failed miserably when fed with test samples that are negative of the images (i.e., convert black to white and vice versa), something that a human would have no issues with. Algorithms are reliable only to the extent of completeness of data used to train them. As always, garbage in implies garbage out.

Further, AI systems have a bias that is inherited from the data that is collected (something that humans consciously try to avoid). For example, a restaurant review system provided poor ratings to Mexican restaurants because the word Mexican is associated with other illegal activities.

Instead of worrying about AI advancements, what if we can come out with a human-machine integrated strategy, including both human and machines, living together in a complex adaptive ecosystem?

Let's call such an ecosystem as human-machine integrated strategy. It mainly includes human cognition and intelligent machines, where human and machines aid each other in problem solving activities. The human-machine integrated strategy for the future is not that of AI-enabled machines replacing humans but of machines and humans existing in a state of symbiosis. The most productive way to utilize AI is to use it to augment human capabilities. Machines do better at specific tasks while humans do better at general tasks. Therefore, a social setting where humans and machines interact while pushing or delegating tasks to each other, if they are not good at it, is an appropriate way to move forward.

The human role in a human-machine integrated architecture is a key differentiator to not only improve the quotient of trust, reciprocity, and likability, but also to allay the fears and concerns associated with proliferation of AI systems.

Intelligent machines blur the lines between computational aspects and interactions with humans for inputs and output. They borrow and extend several established computing principles, like wisdom of crowd computing, collective intelligence computing, social networks, etc. While wisdom of crowd computing focuses on bringing several experts to a central platform, intelligent machines augment decision making of these experts by providing computational intelligence. This hybrid system allows humans and computers to work in harmony toward achieving the common set of goals.

There are several real-life tasks that cannot be accomplished completely by machines. Using human cognition constructively in such tasks can help make problems easier for machines to solve. Researchers have for a long while attempted to make the interaction between humans and machines appear seamless and natural.

Essential ingredients for humans are food, water, energy, and safety. Although these basic objectives seem trivial for a human, one may wonder what they mean for a machine. In case of a machine agent, those needs can be power, network, storage, maintenance, etc. Like in the human world, nothing is free. There has to be a notion of payment for every resource that a machine agent utilizes. It can be imagined as a type of virtual currency, which brings in the notion of reward and penalty for every action that a machine agent takes.

Application Scenarios

HMIS-based applications can primarily have the following different scenarios depending on the composition of human and machine agents, and who bears the responsibility of the actions being taken.

Machine Agents as Human Assistants

Humans are assisted by real-time machine agents to collaborate with diverse multi-cultural agents (sometimes speaking different languages). Machine agents provide necessary information and recommendations to humans, but the final decision is the human's.

For example, consider a machine agent assisting a judge with facts and analysis of previous rulings for similar cases, or a machine agent assisting a recruiter to make a hiring decision based on cultural fit of a candidate.

Fully Autonomous Machine Agents

An autonomous machine agent collaborates with other humans and machine agents. In this scenario, no human is responsible for the machine agent's actions. Hence this scenario is limited to those applications where the risk of the machine agent's actions is very low.

For example, consider an autonomous agent as a virtual figure carrying out tasks in a hazardous physical environment or where the inputs-activities-outputs are clearly defined in a human low-touch environment, or simulating autonomous machine agents as observing, recording, and raising alerts in a quality control process.

A Machine Agent Interacting with Humans

In this scenario, a machine agent is trained to understand and behave according to the preferences and goals of its human counterpart. A machine agent has the ability to negotiate and make decisions for its human. It works like an autonomous machine agent, but it refers back to its human counterpart when in doubt. So the responsibility of actions taken by a machine agent lies with its human counterpart.

For example, consider a machine agent at a call center answering queries from people, being sensitive to their cultural background, or a machine agent performing random A/B testing across the world, taking the cultural background of the people into account.

All Machine Agents Interaction

This is similar to the previous scenario in terms of how agents are trained and who bears the responsibility of actions taken by machine agents. The difference is that there are no humans participating in the interaction. It poses different challenges in terms of how interactions take place among agents.

For example, consider machine agents of decision makers collaborating to come to a consensus on candidates ranking in a hiring process, or machine agents of members of a team selection committee collaboratively choosing a team.

A Human Interacting with Machine Agents

This scenario points to a typical setting where a human walks into a special room in their house or office, where she can be immersed in the virtual environment of another group of people (remotely located) that she wants to collaborate with. All or part of the remote group may be represented by their machine agents.

For example, consider a human who talks to machines agents of friends in a virtual environment to have fun, or a human who conducts an auction, where people send their machine agents to bid.

Governance Framework

Along the lines of the famous three laws of robotics by Isaac Asimov, the HMIS needs to have a governance framework to govern and monitor how human and machine agents carry out their respective tasks in a collaborative fashion. These laws can be used along the three laws of robotics:

- Machine agents will never collect physical features such as skin color, height, weight, etc. as visual input for the purposes of learning or identifying the cultural background of the individuals they cater to. The input will always need to be provided via a formal input channel. This is to make the system impervious to any stereotype associated with physical features.

- A culturally insensitive remark, sentence, vocabulary, or slang will remain tagged insensitive to all cultures, unless its alternative positive aspect is clearly stated for a specific culture.

- Human agents will need an active learning system, capable of incorporating feedback after an interaction, to continuously validate its behavior.

A clear "segregation of responsibility" and event-response matrix needs to be identified for both humans and machine agents. Humans and machine agents can then use the enforced or automated activities to manage interdependencies and interaction. These interdependencies and interactions are either designed a priori or can be executed at runtime based on the scenarios they are exposed to.

Efficiency and specificity are critical and hence coordination rules should be clearly documented by proper management of the interactions. The HMIS should be capable of handling concurrent actions in a multi-agent complex adaptive ecosystem that provides security, reliability, and fault tolerance capabilities. The coordination process should reflect exactly the semantics by which it has been forged. The effect of coordination and accountability should precisely be maintained on the agent interaction space. Safety is another factor in human-machine interaction, which the governance framework should not ignore. As human beings adhere to laws in society, it is as important for the machine agents to abide by a few laws that prevent them from taking drastic measures that may lead to disruptive actions.

Much like in the society we have lawyers and a police force to enforce mandated laws, in HMIS ecosystem we will also need watch dogs for maintaining laws.

For example, in the context of recruitment of human employees in a company through a process of interviews, where final decisions are made by machine agents, the HMIS watch dog will keep a track of whether basic values of equal opportunity are maintained throughout the process or not. For example, if it unknowingly recruits candidates giving rise to gender bias or bias in race, color, or sexual orientation, the HMIS watch dog will not approve such recruitment and will suggest the machine agents reevaluate or change their decisions. Other examples may include HMIS watch dogs to monitor communication texts to detect unacceptable or culturally insensitive words or notions. Further,

HMIS watch dogs may also provide a distributed and autonomous platforms to verify the veracity of communications from machines. Governance of laws is important in HMIS, but equally important is ethics, which should be instilled in agents.

In the business setting, breach of code of conduct should fall in the category of ethics. As described, anything giving rise to racism abusive behavior or even wrongful use of data will fall in the category of ethical breach and, in that case, the punishment should be hefty. One step below ethics comes governance where agents are governed by superior agents in their domain or task. The structure of governance among agents should be hierarchical where there is accountability as well as autonomy among the governing official to take actions based on consensus reached. HMIS watch dogs are expected to become more intelligent through the entire process of governance. This can be done by leveraging adversarial machine learning procedures. As agents need to evolve in their roles and responsibilities, it is important to have enough spam filtering in the agents' learning processes. This can only be done when the HMIS watch dogs become clever by improving filtering mechanisms in the agents' mode of communication, the absence of which may make the agent disrupt the healthy learning process of the agents and in turn the entire ecosystem.

We must also discuss the potential challenges to be encountered on the road to achieving a stable HMIS ecosystem, discussed next.

Trustworthiness and Likability of Machines

The first and foremost requirement to consummate an interaction between two individuals is to accept such an interaction. This acceptance is a function of how trustworthy the individuals see each other while the extent/degree of this acceptance is directly correlated with the likability among the individuals.

This becomes a tricky affair if one of the individuals in context is a machine. While trust and likability do not form a hurdle in a machines' perspective of the human, as it can always be altered programmatically, the vice versa relationship is an uphill battle. With every article or news breakout about an increase in the capability of machines or the highlight of a mistake made by an autonomous machine, the trust of machines takes a beating. The HMIS ecosystem demands a seamless and open exchange of information between humans and machines, which requires the removal of any skepticism from the concerned human's mind regarding the machine. Intellectual superiority does not necessarily make a machine likeable and hence this is a much more complex problem than it may seem on paper. For example, machines that make mistakes seem to make humans more comfortable as compared to machines that are always correct.

Human-Machine Relationship

The human-machine relationship is a symbiotic one where both entities are dependent on each other. While machines are superior to humans in performing well defined tasks, humans are superior in dynamic tasks that are not fully controlled and are affected by uncertain factors. Since real-life situations can consist of both types of tasks, the degree of freedom needs to be divided among humans and machines.

The HMIS ecosystem consists of an intermingling of the network of machines and the network of humans. As machines grow more and more autonomous, it is important to maintain a check on the nature of their relationship and the changes in the dependency paradigm.

Defining the relationship is crucial; if an autonomous agent commits a mistake or breaks a law, this relationship would be crucial to identify the entity that would take up responsibility. Challenges like the dynamics of autonomy, ownership of tasks, and distribution of work must be addressed.

All Humans Are Different

As humans, one interacts with different people differently. The way people interact with their families is different from the way they interact with their colleagues, as there is a high dimensional context that goes into setting the tone and level of the interaction. The context does not limit itself to the relationship shared among the interacting parties, but also extends to the individual characteristics of the parties involved, such as age, gender, etc. Naturally, on the road to making machines more and more human-like, identifying the maximum context is crucial and hence becomes our next challenge.

While a good amount of research has been done in identifying physical characteristics of humans, such as gender and age, there is a reasonable amount of work being done in identifying dynamic characteristics such as mood, emotions, etc. from tone, expressions, and body language. An interesting aspect and an added complexity in using this identified context is that it is culture specific. Even when it comes to the interaction between a machine and a human, culture plays an important role in how machines are perceived differently by individuals of different cultural backgrounds. While machines need to understand what certain gestures mean in different cultures, they also need to understand how their responses will be received in the same context as well.

People and technology must play their roles, as depicted in Figure 11-1, and humans must constantly evolve the design of the machines, provide governance frameworks, monitor capabilities, and embrace the notion that an integrated human-machine strategy will not eliminate them but will augment their efforts.

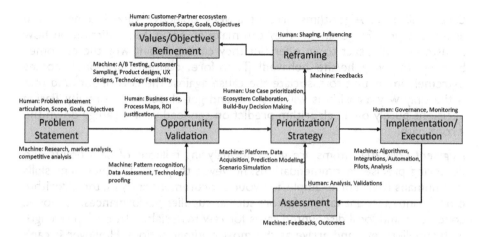

Figure 11-1. A human-machine integrated strategy map

Conclusion

The rate at which data is growing is far outpacing our ability to extract business value from it. Most raw data to insights solutions are designed for volume. Few are designed for complexity. Fewer still excel at both volume and complexity. Thus, the challenge is how to extract business value from massively complex data.

At the same time, there are companies making audacious attempts to such as "mind as a service". Nectome[7] is researching to develop a sophisticated brain banking technology to digitize a clinically preserved brain and use that information to recreate the person's mind. Leveraging the synapses between neurons in a brain and reverse-engineering the brain, it is possible to keep all its memories intact. The result? Right from the childhood through all the highs and lows of a person's life, the experiences associated with those events, the lessons and the wisdom, all can be made available, later on, as a service. How close are we to this possibility, though, only time will tell.

For many, the word "algorithm" paints a complex incomprehensible set of mathematical formula. However, the reality is that algorithms are all around us, mostly working in an invisible mode but delivering significant business outcomes as well as uplifting our quality of life. No doubt, there are massive advantages to using algorithms within a business context; however, there are also a few challenges that need to be addressed.

---

[7]https://nectome.com/

Contextualization: Algorithms present a very objective view leading to the final prediction. For example, they can make an accurate prediction on how a customer will react to an offer, but they can't pinpoint why the customer behaved the way he/she behaved! Therefore, contextualization becomes extremely important to associate the cause against the correlation, and that is the only way algorithms will improve and get closer to the reality. Basing decisions purely on an algorithm prediction can lead your strategy down the wrong path.

Judgment skills: Algorithms are good at analyzing millions of data points and delivering precise recommendations; however, they lack the judgment skills that humans have. For example, for your procurement analysis, the algorithm can take into account various factors such as supplier performances, economic conditions, and local discounted rates for raw materials, relationship strength with suppliers, etc. and arrive at the most optimal options. However, it can't negotiate with the supplier the way an experienced human would to get the best rates.

During our numerous conversations with business owners, technology practitioners, and CXOs, one question was oft repeated: "Do you think algorithms will impact our organization and our roles?" To answer, we borrow few stats from a Gartner report,[8] which estimates that 20% of all business content will come from machines by 2018; while autonomous software agents will participate in 5% of all economic transactions by 2020.

To summarize, given the abundance of data sources (thanks to digital becoming a way of life and businesses) and fast-paced technology evolution (IoT, cloud, automation, blockchain, and ML/DL), algorithmic business is going to become the norm, not an aberration. While the cost of prediction will get cheaper and cheaper, the need for judgment skills will rise and the human element cannot be taken out altogether. Organizations will need to rethink their strategies and focus on products, services, customers, markets and human assets. They also need to add one more dimension—human-machine integrated AI strategy—into their enterprise game plans.

References

1. http://www.digitalistmag.com/future-of-work/2017/07/19/automation-will-lead-to-collaboration-between-man-machine-05217208

2. http://www.worktechacademy.com/changes-everything-working-human-robotic-ecosystem/

---

[8]https://www.gartner.com/smarterwithgartner/gartner-predicts-our-digital-future/

3. https://www2.deloitte.com/insights/us/
   en/deloitte-review/issue-20/augmented-
   intelligence-human-computer-collaboration.
   html

4. https://blog.dominodatalab.com/ai-enterprise/

5. https://dataorigami.net/blogs/napkin-
   folding/17543555-datas-use-in-the-21st-
   century

6. https://towardsdatascience.com/the-pac-
   framework-how-non-technical-executives-
   should-think-about-artificial-intelligence-
   b2d733036a52

7. https://www.linkedin.com/pulse/cognitive-
   collaboration-why-humans-computers-think-
   better-lewis/

Index

Printed in the United States
By Bookmasters